Mathematical Engineering

More information about this series at http://www.springer.com/series/8445

Mass Per Pettersson • Gianluca Iaccarino
Jan Nordström

Polynomial Chaos Methods for Hyperbolic Partial Differential Equations

Numerical Techniques for Fluid Dynamics Problems in the Presence of Uncertainties

Springer

Mass Per Pettersson
Uni Research
Bergen, Norway

Jan Nordström
Department of Mathematics
Computational Mathematics
Linköping University
Sweden

Gianluca Iaccarino
Department of Mechanical Engineering
and Institute for Computational
and Mathematical Engineering
Stanford University, USA

Additional material to this book can be downloaded from http://extras.springer.com.

ISSN 2192-4732 ISSN 2192-4740 (electronic)
Mathematical Engineering
ISBN 978-3-319-10713-4 ISBN 978-3-319-10714-1 (eBook)
DOI 10.1007/978-3-319-10714-1

Library of Congress Control Number: 2014950671

Springer Cham Heidelberg New York Dordrecht London

Printed on acid-free paper

Springer International Publishing AG Switzerland is part of Springer Science+Business Media (www.springer.com)

Preface

Uncertainty quantification in computational physics is a broad research field that has spurred increasing interest in the last two decades, partly due to the growth of computer power. The objective of this textbook is the analysis and design of numerical techniques for solving equations representing conservation laws subject to uncertainty. In particular, the focus is on stochastic Galerkin methods that require non-trivial development of new numerical solvers for hyperbolic and mixed type problems. There are already textbooks covering the stochastic Galerkin and other polynomial chaos methods from a general perspective, cf. [1–3]: this textbook is more specialized in its scope. To enhance understanding of the material presented, we provide exercises and code scripts and building blocks that can be extended to new problem settings.

The interest in stochastic Galerkin methods has burgeoned because of the availability of ever more powerful computers that can handle the computational cost inherent to large system implementations. Moreover, these methods have positive numerical properties that make them attractive for handling complex situations. Specifically, the mathematical formulation leads to systems of equations that resemble the original conservation laws, allowing us to make extensive use of available numerical analysis tools and techniques. At the same time, the stochastic Galerkin method is an attractive alternative for complex problems involving partial differential equations and multiple uncertain variables (herein referred to as stochastic dimensions).

Chapters 1–3 introduce and give a brief overview of the basic concepts of uncertainty quantification and the stochastic Galerkin method. Chapter 4 is devoted to spatial discretization methods for conservation laws under uncertainty. In particular, we introduce the so-called SBP-SAT finite difference technique based on summation-by-parts operators (SBP) and weak boundary conditions using simultaneous approximation terms (SAT). The SBP-SAT schemes allow for the design of stable high-order accurate schemes. Summation by parts is the discrete equivalent of integration by parts and the matrix operators that are presented lead to energy estimates that, in turn, lead to provable stability in combination with the SAT terms. The semidiscrete stability follows naturally from the continuous analysis of well-posedness which provides the boundary conditions in the SBP-SAT technique. Chapters 5–9 present in-depth analysis of linear and nonlinear stochastic Galerkin conservation laws, complemented by exercises and scripts. We provide the reader with computer codes for solving the advection-diffusion equation and the inviscid Burgers' equation with the stochastic Galerkin method. These codes can also be used as templates for extension to more complex problems.

This textbook is intended for an audience with some prior knowledge of uncertainty quantification. Basic concepts of probability theory, statistics and numerical analysis are also assumed to be familiar to the reader. For a more general exposition and further details on the basic concepts, we refer to the existing literature in the field.

This textbook has benefited from numerous collaborations and discussions with Alireza Doostan (who co-authored the material contained in Chap. 5), Antony Jameson, Xiangyu Hu, Rémi Abgrall and Paul Constantine. We would like to thank Margot Gerritsen for constructive feedback and suggestions for improvement. Financial support was partially provided by KAUST under the Stanford/KAUST Academic Excellence Alliance (AEA) collaboration (UDGIA Award 48803). Gianluca Iaccarino wishes to thank the *Borrister crew* for support in completing the final revision of the text.

Bergen, Norway — Mass Per Pettersson
Stanford, USA — Gianluca Iaccarino
Linköping, Sweden — Jan Nordström

References

1. Xiu D (2010) Numerical methods for stochastic computations: a spectral method approach. Princeton University Press, Princeton. http://www.worldcat.org/isbn/9780691142128
2. Le Maître OP, Knio OM (2010) Spectral methods for uncertainty quantification, 1st edn. Springer, Dordrecht/New York
3. Smith R (2013) Uncertainty quantification: theory, implementation, and applications. Computational science and engineering. SIAM, Philadelphia

Contents

Acronyms

CFL	Courant-Friedrichs-Lewy
ENO	Essentially Non-Oscillatory Scheme
gPC	Generalized polynomial chaos
HLL	Harten-Lax; van Leer
IBVP	Initial-boundary value problem
KL	Karhunen-Loève
ME-gPC	Multi-element generalized polynomial chaos
MUSCL	Monotone upstream-centered scheme for conservation laws
MW	Multiwavelet
ODE	Ordinary differential equation
PC	Polynomial chaos
PDE	Partial differential equation
SAT	Simultaneous approximation term
SBP	Summation by parts
TVD	Total variation diminishing
UQ	Uncertainty quantification

Part I
Introductory Concepts and Background

Chapter 1
Introduction

In many physical problems, knowledge is limited in quality and quantity by variability, bias in the measurements and limitations in the measurements: these are all sources of uncertainties. When we attempt to solve the problem numerically, we must account for those limitations, and in addition, we must identify possible shortcomings in the numerical techniques employed. Incomplete understanding of the physical processes involved will add to the sources of possible uncertainty in the models employed. In a general sense, we distinguish between *errors* and *uncertainty* simply by saying that errors are *recognizable deficiencies not due to lack of knowledge*, whereas uncertainties are *potential* and directly related to lack of knowledge [1]. This definition clearly identifies errors as deterministic quantities and uncertainties as stochastic in nature; uncertainty estimation and quantification are, therefore, typically treated within a probabilistic framework.

Uncertainty quantification is also a fundamental step towards validation and certification of numerical methods to be used for critical decisions. Fields of application of uncertainty quantification include, but are not limited to, turbulence, climatology [18], turbulent combustion [19], flow in porous media [5, 6], fluid mixing [26] and computational electromagnetics [4].

An example of the need for uncertainty quantification in applications related to the methods and problems studied here is the investigation of the aerodynamic stability properties of an airfoil. Uncertainties in physical parameters such as structural frequency and initial pitch angle affect the limit cycle oscillations. One approach in particular, the *polynomial chaos* method, has been used to obtain a statistical characterization of the stability limits and to calculate the risk for system failure [2, 25]; this approach will be studied in detail throughout this monograph.

The sources of uncertainty that we consider here are imprecise knowledge of the input data, e.g., uncertainty due to limited observations or measurement errors. This imprecision results in numerical models that are subject to uncertainty in boundary or initial conditions, in model parameter values and even in the geometry of the

M.P. Pettersson et al., *Polynomial Chaos Methods for Hyperbolic Partial Differential Equations*, Mathematical Engineering,
DOI 10.1007/978-3-319-10714-1_1

physical domain of the problem (input uncertainty). Uncertainty quantification in the sense used here is concerned with the propagation of input uncertainty through the numerical model in order to clearly identify and quantify the uncertainty in the output quantities of interest.

Without going into detail how to transform a set of data into probability distributions of the input variables [7], the starting point will be a partial differential equation formulation where parameters and initial and boundary conditions are uncertain but determined in terms of probability distributions. Random variables are used to parametrize the uncertainty in the input data. A spectral series representation, the generalized chaos series expansion, is then used to represent the solution to the problem of interest.

The test problems that will be investigated here are evidently subject to modeling error, were we to use them as representative models of real-world phenomena. For instance, we disregard viscous forces in many of the flow problems, and focus only on one-dimensional physical situations. Thus, we do not account for uncertainty in the physical and mathematical models themselves. In real-world problems, this omission would be an important point. If the conceptual model is erroneous, for instance due to an incompressibility assumption for a case of high Mach number flow, then there is very little use for its solution, no matter the degree of accuracy of the representation of variability in the input parameters [20].

Of the several approaches to propagate the input uncertainty in numerical computations, the simplest one is the Monte Carlo method where a vast number of simulations are performed to compute the output statistics. Conversely, in the polynomial chaos approach, the solution is expressed as a truncated series and only one simulation is performed. The dimension of the resulting system of equations grows with the number of terms retained in the series (the order of the polynomial chaos expansion) and the dimension of the stochastic input.

An increased number of Monte Carlo simulations implies a solution with better converged statistics; on the other hand, in the polynomial chaos approach, one single simulation is sufficient to obtain a complete statistical characterization of the solution. However, the accuracy of this solution is dependent on the order of polynomials considered, and therefore on the truncation in the chaos expansion. Also, for optimal convergence the polynomial chaos solution must be smooth with respect to the parameters describing the input uncertainty [24].

1.1 Theory for Initial Boundary Value Problems

Throughout this book, the Uncertainty Quantification (UQ) problem at hand is governed by an Initial-Boundary-Value Problem (IBVP) and the main part of the general theory will be reviewed in short here. The material covered in this section can be found in [3, 8, 9, 12–15, 17, 21–23].

1.1.1 The Continuous Problem

Consider the initial-boundary-value problem

$$\begin{aligned} \boldsymbol{u}_t + \boldsymbol{P}(x,t,\partial_x)\boldsymbol{u} &= \boldsymbol{F}(x,t), \quad 0 \le x \le 1, \quad t \ge 0, \\ \boldsymbol{u}(x,0) &= \boldsymbol{f}(x), \\ \boldsymbol{L}_0(t,\partial_x)\boldsymbol{u}(0,t) &= \boldsymbol{g}_0(t), \\ \boldsymbol{L}_0(t,\partial_x)\boldsymbol{u}(1,t) &= \boldsymbol{g}_1(t), \end{aligned} \tag{1.1}$$

where $\boldsymbol{u} = (u_1, \ldots, u_n)^T$ is the solution vector and $\boldsymbol{P}$ is a differential operator with smooth matrix coefficients. $\boldsymbol{L}_0$ and $\boldsymbol{L}_1$ are differential operators defining the boundary conditions. The boundary data of the problem are $\boldsymbol{g}_0(t), \boldsymbol{g}_1(t)$, the initial data are $\boldsymbol{f}(x)$, and $\boldsymbol{F}(x,t)$ is a forcing function.

Definition 1.1. The IBVP (1.1) with $\boldsymbol{F} = \boldsymbol{g}_0 = \boldsymbol{g}_1 = \boldsymbol{0}$ is well-posed, if for every $\boldsymbol{f} \in C^\infty$ that vanishes in a neighborhood of $x = 0, 1$, it has a unique smooth solution that satisfies the estimate

$$\|\boldsymbol{u}(\cdot,t)\| \le K e^{\alpha_c t} \|\boldsymbol{f}\|, \tag{1.2}$$

where K, α_c are constants independent of $\boldsymbol{f}$. The estimate (1.2) must be obtained by using a minimal number of boundary conditions.

A stronger and more practical, albeit more difficult to prove, version of well-posedness, including nonzero boundary data and forcing function, is given by

Definition 1.2. The IBVP (1.1) is *strongly well-posed*, if it is well-posed and

$$\|\boldsymbol{u}(\cdot,t)\|^2 \le K(t)\left(\|\boldsymbol{f}\|^2 + \int_0^t \left(\|\boldsymbol{F}(\cdot,\tau)\|^2 + |\boldsymbol{g}_0(\tau)|^2 + |\boldsymbol{g}_1(\tau)|^2\right) d\tau\right) \tag{1.3}$$

holds. The function $K(t)$ is bounded for every finite time and is independent of $\boldsymbol{F}, \boldsymbol{g}_0, \boldsymbol{g}_1, \boldsymbol{f}$.

The boundary and initial data are compatible in the definitions above, as is necessary in order to ensure a smooth solution. Compatibility means that the initial condition at the boundaries must be consistent with the boundary conditions at the initial time. More details on compatibility can be found in [8].

As an example of the relevance of having estimates like (1.3), we consider a perturbed version of problem (1.1) with data $\boldsymbol{F} + \delta\boldsymbol{F}, \boldsymbol{g}_0 + \delta\boldsymbol{g}_0, \boldsymbol{g}_1 + \delta\boldsymbol{g}_1, \boldsymbol{f} + \delta\boldsymbol{f}$ and solution $\boldsymbol{v}$. Assuming $\boldsymbol{P}$ in (1.1) to be a linear operator, we obtain a similar problem for $\boldsymbol{v} - \boldsymbol{u}$ by subtracting the IBVP for $\boldsymbol{u}$ from the IBVP for $\boldsymbol{v}$. The corresponding data are the perturbed values $\delta\boldsymbol{F}, \delta\boldsymbol{g}_0, \delta\boldsymbol{g}_1, \delta\boldsymbol{f}$. Clearly, the estimate (1.3) now states that the difference $\boldsymbol{v} - \boldsymbol{u}$ is small for small differences in data.

1.1.2 The Semidiscrete Problem

We keep time continuous and discretize (1.1) in space. Semidiscretization results in a system of ordinary differential functions (ODEs) that is easier to analyze than the fully discrete problem. Let $x_j = jh$, $j = 1, \ldots, m$ where $h = 1/(m-1)$ is the grid spacing. We define the grid functions $\boldsymbol{f}_j = \boldsymbol{f}(x_j)$ and $\boldsymbol{F}_j(t) = \boldsymbol{F}(x_j, t)$ and associate the approximate solution $\boldsymbol{v}_j(t)$ to each grid point. We form vectors of the grid functions as $\vec{\boldsymbol{v}} = (\boldsymbol{v}_1, \boldsymbol{v}_2, \ldots, \boldsymbol{v}_m)^T$, $\vec{\boldsymbol{f}} = (\boldsymbol{f}_1, \boldsymbol{f}_2, \ldots, \boldsymbol{f}_m)^T$ and $\vec{\boldsymbol{F}} = (\boldsymbol{F}_1, \boldsymbol{F}_2, \ldots, \boldsymbol{F}_m)^T$ and use the notion *smooth grid function* to denote a grid function being the projection of a smooth function. Furthermore, we use $\|\cdot\|_h$ to denote a discrete L^2-equivalent norm.

We approximate (1.1) by

$$\vec{\boldsymbol{v}}_t + \tilde{\boldsymbol{P}}(\vec{\boldsymbol{x}}, t)\vec{\boldsymbol{v}} = \vec{\boldsymbol{F}} + \vec{\boldsymbol{S}}, \quad t \geq 0$$
$$\vec{\boldsymbol{v}}(0) = \vec{\boldsymbol{f}}, \tag{1.4}$$

where $\tilde{\boldsymbol{P}}$ is the discrete approximation of $\boldsymbol{P}$. $\vec{\boldsymbol{S}} = \vec{\boldsymbol{S}}(\boldsymbol{g}_0, \boldsymbol{g}_1)$ is the so-called *simultaneous approximation term* (SAT) which implements the boundary conditions weakly (see [3]). The SAT term, which is one part of the so-called summation-by-parts simultaneous approximation term (SBP-SAT) technique (see [23] for a review), will be discussed extensively later. $\vec{\boldsymbol{S}}$ is zero except at a few points close to the boundaries. The next definition is in analogy with Definition 1.1 above.

Definition 1.3. Consider (1.4) with $\vec{\boldsymbol{F}} = \boldsymbol{0}$, $\boldsymbol{g}_0 = \boldsymbol{g}_1 = \boldsymbol{0}$. Let $\vec{\boldsymbol{f}}$ be the projection of a C^∞ function that vanishes at the boundaries. The approximation is *stable* if, for all $h \leq h_0$,

$$\|\vec{\boldsymbol{v}}(t)\|_h \leq K e^{\alpha_d t} \|\vec{\boldsymbol{f}}\|_h \tag{1.5}$$

holds and K, α_d, h_0 are constants independent of $\vec{\boldsymbol{f}}$.

The following definition corresponds to Definition 1.2 and allows for nonzero boundary data and forcing function.

Definition 1.4. The approximation (1.4) is *strongly stable* if it is stable and

$$\|\vec{\boldsymbol{v}}(t)\|_h^2 \leq K(t)\left(\|\vec{\boldsymbol{f}}\|_h^2 + \max_{\tau\in[0,t]} \|\vec{\boldsymbol{F}}(\tau)\|_h^2 + \max_{\tau\in[0,t]} \|\boldsymbol{g}_0(\tau)\|_h^2 + \max_{\tau\in[0,t]} \|\boldsymbol{g}_1(\tau)\|_h^2\right) \tag{1.6}$$

holds. $K(t)$ is bounded for any finite t and is independent of $\vec{\boldsymbol{F}}, \boldsymbol{g}_0, \boldsymbol{g}_1, \vec{\boldsymbol{f}}$.

The relevance of having estimates like (1.6) is similar to the relevance of having the estimate (1.3) that was discussed above in the continuous section. An identical exercise on a linear perturbed problem shows that the estimate (1.6) guarantees that the difference between two solutions is small for small differences in data.

Although the definitions of (strong) well-posedness and (strong) stability are similar, the bounds in the corresponding estimates need not be the same, (see [8, 13–15]). In all the above definitions, the schemes are semi-discrete, i.e., time is left continuous. Clearly, only fully discrete schemes are useful in practice. In [10], it was shown that semi-discrete stable schemes are, under certain conditions, stable when discretized in time using Runge-Kutta schemes. Recently it was shown in [11, 16] how to extend the SBP-SAT technique in space to the time-domain, where fully discrete sharp energy estimates are obtained.

We will later use the notion of *energy stability*, by which we mean that (i) the continuous problem has boundary conditions that lead to an energy estimate, and (ii) the numerical scheme leads to a corresponding discrete energy estimate. The SAT terms take care of (ii) if (i) is satisfied. The procedure is almost automatic when SBP-SAT schemes are used (see [23] for details).

Finally, some remarks are offered on well-posedness and stability for non-linear problems. The status of the theory is not satisfactory. The estimates and bounds on the solution as shown in the definitions above are clearly valuable, inasmuch as they prevent blow-up of the solution. However, the existence of the bounds does not necessarily imply well-posedness since an equation for the difference between two solutions is non-trivial to obtain. However, this difficulty may in some cases be purely technical. Also, if one knows that the non-linear solution is reasonably smooth, one can use the *linearization and localization principles* formulated in [9] and arrive at well-posedness. In the rest of this book we will not go into the uncharted territory of non-linear theory for IBVPs but will rely on the *linearization and localization principles* when checking well-posedness and stability.

1.2 Outline

The aim in Chaps. 2 and 3 is to lay a theoretical background for the numerical and theoretical results to be presented in subsequent chapters. The theory of spectral expansions of random fields is outlined in Chap. 2, followed by an exposition of methods for the solution of PDEs with stochastic input in Chap. 3. Numerical discretization schemes are described in Chap. 4. As motivation for the use of generalized polynomial chaos methods as well as the numerical methods of our choice, Chap. 5 introduces an advection-diffusion problem with a smooth solution. For general nonlinear conservation laws, the solutions are non-smooth in the deterministic case. In order to find suitable numerical methods and robust stochastic representations for the corresponding stochastic Galerkin formulations, we analyze the regularity of conservation laws with stochastic input conditions. In Chap. 6, we investigate Burgers' equation with uncertain boundary conditions in terms of regularity. This chapter illustrates the method of imposition of weak characteristic boundary conditions employed in all subsequent chapters. Next, we investigate Burgers' equation in terms of the effect of incomplete boundary conditions in Chap. 7. A stochastic Galerkin method for the Euler equations combining robust representation of input uncertainty with shock-capturing methods is presented in Chap. 8. Finally, in Chap. 9, we generalize the analysis of regularity to a two-phase flow problem. Based on the spatial localization of smooth and non-smooth solution regions, we then combine high-order methods with shock-capturing methods into a hybrid scheme.

References

1. AIAA (American Institute of Aeronautics and Astronautics) Staff (1998) AIAA Guide for the Verification and Validation of Computational Fluid Dynamics Simulations. American Institute of Aeronautics & Astronautics, Reston, VA
2. Beran PS, Pettit CL, Millman DR (2006) Uncertainty quantification of limit-cycle oscillations. J Comput Phys 217(1):217–247. doi:http://dx.doi.org/10.1016/j.jcp.2006.03.038
3. Carpenter MH, Nordström J, Gottlieb D (1999) A stable and conservative interface treatment of arbitrary spatial accuracy. J Comput Phys 148(2):341–365. doi:http://dx.doi.org/10.1006/jcph.1998.6114
4. Chauvière C, Hesthaven JS, Lurati L (2006) Computational modeling of uncertainty in time-domain electromagnetics. SIAM J Sci Comput 28(2):751–775. doi:http://dx.doi.org/10.1137/040621673
5. Christie M, Demyanov V, Erbas D (2006) Uncertainty quantification for porous media flows. J Comput Phys 217(1):143–158. doi:http://dx.doi.org/10.1016/j.jcp.2006.01.026
6. Ghanem RG, Dham S (1998) Stochastic finite element analysis for multiphase flow in heterogeneous porous media. Porous Media 32:239–262
7. Ghanem RG, Doostan A (2006) On the construction and analysis of stochastic models: characterization and propagation of the errors associated with limited data. J Comput Phys 217(1):63–81. doi:http://dx.doi.org/10.1016/j.jcp.2006.01.037
8. Gustafsson B, Kreiss HO, Oliger J (1995) Time dependent problems and difference methods, 1st edn. Wiley, New York
9. Kreiss HO, Lorenz J (1989) Initial boundary value problems and the Navier–Stokes equations. Academic, New York
10. Kreiss HO, Wu L (1993) On the stability definition of difference approximations for the initial boundary value problem. Appl Numer Math 12:213–227
11. Lundquist T, Nordström J (2014) The sbp-sat technique for initial value problems. J Comput Phys 270:86–104
12. Nordström J (1995) The use of characteristic boundary conditions for the Navier-Stokes equations. Comput Fluids 24(5):609–623
13. Nordström J (2006) Conservative finite difference formulations, variable coefficients, energy estimates and artificial dissipation. J Sci Comput 29(3):375–404. doi:http://dx.doi.org/10.1007/s10915-005-9013-4
14. Nordström J (2007) Error bounded schemes for time-dependent hyperbolic problems. SIAM J Sci Comp 30(1):46–59. doi:10.1137/060654943
15. Nordström J, Eriksson S, Eliasson P (2012) Weak and strong wall boundary procedures and convergence to steady-state of the Navier-Stokes equations. J Comput Phys 231(14):4867–4884
16. Nordström J, Lundquist T (2013) Summation-by-parts in time. J Comput Phys 251:487–499
17. Nordström J, Svärd M (2005) Well-posed boundary conditions for the Navier-Stokes equations. SIAM J Numer Anal 43(3):1231–1255
18. Poroseva S, Letschert J, Hussaini MY (2005) Uncertainty quantification in hurricane path forecasts using evidence theory. In: APS meeting abstracts, pp B1+, Chicago, IL

19. Reagan MT, Najm HN, Ghanem RG, Knio OM (2003) Uncertainty quantification in reacting-flow simulations through non-intrusive spectral projection. Combust Flame 132(3):545–555
20. Roache PJ (1997) Quantification of uncertainty in computational fluid dynamics. Annu Rev Fluid Mech 29:123–160. doi:10.1146/annurev.fluid.29.1.123
21. Strikwerda JC (1977) Initial boundary value problems for incompletely parabolic systems. Commun Pure Appl Math 9(3):797–822
22. Svärd M, Nordström J (2006) On the order of accuracy for difference approximations of initial-boundary value problems. J Comput Phys 218(1):333–352. doi:http://dx.doi.org/10.1016/j.jcp.2006.02.014
23. Svärd M, Nordström J (2014) Review of summation-by-parts schemes for initial-boundary-value problems. J Comput Phys 268:17–38
24. Wan X, Karniadakis GE (2006) Long-term behavior of polynomial chaos in stochastic flow simulations. Comput Methods Appl Math Eng 195:5582–5596
25. Witteveen JAS, Sarkar S, Bijl H (2007) Modeling physical uncertainties in dynamic stall induced fluid-structure interaction of turbine blades using arbitrary polynomial chaos. Comput Struct 85(11–14):866–878. doi:http://dx.doi.org/10.1016/j.compstruc.2007.01.004
26. Yu Y, Zhao M, Lee T, Pestieau N, Bo W, Glimm J, Grove JW (2006) Uncertainty quantification for chaotic computational fluid dynamics. J Comput Phys 217(1):200–216. doi:http://dx.doi.org/10.1016/j.jcp.2006.03.030

Chapter 2
Random Field Representation

Nonlinear conservation laws subject to uncertainty are expected to develop solutions that are discontinuous in spatial as well as in stochastic dimensions. In order to allow piecewise continuous solutions to the problems of interest, we follow [7] and broaden the concept of solutions to the class of functions equivalent to a function f, denoted $\mathscr{C}_f$, and define a normed space that does not require its elements to be smooth functions. Let $(\Omega, \mathscr{F}, \mathscr{P})$ be a probability space with event space Ω, and probability measure $\mathscr{P}$ defined on the σ-field $\mathscr{F}$ of subsets of Ω. Let $\boldsymbol{\xi} = \{\xi_j(\omega)\}_{j=1}^N$ be a set of N independent and identically distributed random variables for $\omega \in \Omega$. We consider *second-order random fields*, i.e., we consider f belonging to the space

$$L^2(\Omega, \mathscr{P}) = \left\{ C_f | f \text{ measurable w.r.t.} \mathscr{P}; \int_\Omega f^2 d\mathscr{P}(\xi) < \infty \right\}. \tag{2.1}$$

The inner product between two functionals $a(\xi)$ and $b(\xi)$ belonging to $L^2(\Omega, \mathscr{P})$ is defined by

$$\langle a(\xi) b(\xi) \rangle = \int_\Omega a(\xi) b(\xi) d\mathscr{P}(\xi). \tag{2.2}$$

This inner product induces the norm $\|f\|^2_{L_2(\Omega,\mathscr{P})} = \langle f^2 \rangle$.

Spectral representations of random functionals aim at finding a series expansion in the form

$$f(\xi) = \sum_{k=0}^{\infty} f_k \psi_k(\xi(\omega)),$$

where $\{\psi_k(\xi)\}_{k=0}^\infty$ is the set of basis functions and $\{f_k\}_{k=0}^\infty$ is the set of coefficients to be determined.

M.P. Pettersson et al., *Polynomial Chaos Methods for Hyperbolic Partial Differential Equations*, Mathematical Engineering,
DOI 10.1007/978-3-319-10714-1_2

The coefficients are defined by the *projections*

$$f_k = \langle \psi_k f \rangle, \quad k = 0, 1, \ldots$$

2.1 Karhunen-Loève Expansion

The Karhunen-Loève expansion [10, 14] provides a series representation of a random field in terms of its spatial correlation (covariance kernel). Any second-order random field $f(\boldsymbol{x}, \omega)$ on a spatial domain $\boldsymbol{\Omega}_x$ can be represented as the Karhunen-Loève expansion

$$\boldsymbol{f}(\boldsymbol{x}, \omega) = \bar{\boldsymbol{f}}(\boldsymbol{x}) + \sum_{k=1}^{\infty} \eta_k(\omega) \sqrt{\lambda_k} \boldsymbol{\phi}_k^{KL}(\boldsymbol{x}),$$

where $\bar{\boldsymbol{f}}(\boldsymbol{x})$ is the mean of $\boldsymbol{f}(\boldsymbol{x}, \omega)$, the random variables η_k are uncorrelated with mean zero, and λ_k and $\boldsymbol{\phi}_k^{KL}$ are the eigenvalues and eigenfunctions of the covariance kernel, respectively.

The generalized eigenpairs $(\lambda_k, \boldsymbol{\phi}_k^{KL})$ can be determined from the solution of the generalized eigenvalue problem

$$\int_{\Omega_x} \boldsymbol{C}_f(\boldsymbol{x}, \boldsymbol{x}') \boldsymbol{\phi}_k^{KL}(\boldsymbol{x}') d\boldsymbol{x}' = \lambda_k \boldsymbol{\phi}_k^{KL}(\boldsymbol{x}), \quad k \in \mathbb{N}^+, \tag{2.3}$$

where the *covariance function* $\boldsymbol{C}_f$ defines the two-point spatial statistics. The covariance function $\boldsymbol{C}_f$ does not contain information sufficient to determine the joint probability distribution of the random variables $\{\eta_k\}$. Instead, the joint probability of these random variables must be determined by data.

The Karhunen-Loève expansion is bi-orthogonal, i.e.,

$$\left\langle \boldsymbol{\phi}_j^{KL}(\boldsymbol{x}), \boldsymbol{\phi}_k^{KL}(\boldsymbol{x}) \right\rangle_{\Omega_x} \equiv \int_{\Omega_x} \left(\boldsymbol{\phi}_j^{KL}(\boldsymbol{x}) \right)^T \boldsymbol{\phi}_k^{KL}(\boldsymbol{x}) d\boldsymbol{x} = \delta_{jk}, \tag{2.4}$$

$$\left\langle \eta_j \eta_k \right\rangle_{\Omega} \equiv \int_{\Omega} \eta_j \eta_k d\mathscr{P} = \delta_{jk}. \tag{2.5}$$

For random fields with known covariance structure, the Karhunen-Loève expansion is optimal in the sense that it minimizes the mean-squared error. The covariance function of the output of a problem is in general not known a priori. However, Karhunen-Loève representations of the input data can often be combined with generalized chaos expansions, presented in the next section.

2.2 Generalized Chaos Expansions

Infinite series expansions in terms of functions that are orthogonal with respect to the probability measure of some random parametrization are used for representation of stochastic quantities of interest. The corresponding series expansions of these basis functions are referred to as generalized chaos expansions. Possible choices include polynomials and wavelets.

2.2.1 Generalized Polynomial Chaos Expansion

The polynomial chaos (PC) framework based on series expansions of Hermite polynomials of Gaussian random variables was introduced by Ghanem and Spanos [9] and builds on the theory of homogeneous chaos introduced by Wiener in 1938 [18]. Any second-order random field can be expanded as a generalized Fourier series in the set of orthogonal Hermite polynomials, which constitutes a complete basis in the Hilbert space $L^2(\Omega, \mathscr{P})$ defined by (2.1). The resulting polynomial chaos series converges in the $L^2(\Omega, \mathscr{P})$ sense as a consequence of the Cameron-Martin theorem [3]. Although not limited to represent functions with Gaussian distribution, the polynomial chaos expansion achieves the highest convergence rate for Gaussian functions. Xiu and Karniadakis [20] introduced the *generalized polynomial chaos* (gPC) expansion, where random functions are represented by any set of hypergeometric polynomials from the Askey scheme [2]. Hence, a function with uniform distribution is optimally represented by Legendre polynomials that are orthogonal with respect to the uniform measure, and a gamma-distributed input by Laguerre polynomials that are orthogonal with respect to the gamma measure, and so on. The optimality of the choice of stochastic expansion pertains to the representation of the input; the representation of the output of a nonlinear problem will likely be highly nonlinear as expressed in the basis of the input.

The Cameron-Martin theorem applies also to gPC with non-Gaussian random variables, but only when the probability measure $\mathscr{P}(\xi)$ of the stochastic expansion variable ξ is uniquely determined by the sequence of moments,

$$\langle \xi^k \rangle = \int_\Omega \xi^k d\mathscr{P}(\xi), \quad k \in \mathbb{N}_0.$$

This is not always the case in situations commonly encountered; for instance, the lognormal generalized chaos does not satisfy this property. Thus, there are cases when the gPC expansion does not converge to the true limit of the random variable under expansion [6]. However, lognormal random variables may be successfully represented by gPC satisfying the determinacy of moments (cf. [6] for a detailed exposition on this topic), e.g., Hermite polynomial chaos expansion. This motivates our choice to use Hermite polynomial chaos expansion to represent lognormal viscosity in Chap. 5.

Consider a generalized chaos basis $\{\psi_i(\xi)\}_{i=0}^{\infty}$ spanning the space of second-order (i.e., finite variance) random processes on this probability space. The basis functionals are assumed to be orthonormal, i.e., they satisfy

$$\langle \psi_i \psi_j \rangle = \delta_{ij}. \tag{2.6}$$

Any second-order random field $u(x,t,\xi)$ can be expressed as

$$u(x,t,\xi) = \sum_{i=0}^{\infty} u_i(x,t)\psi_i(\xi), \tag{2.7}$$

where the coefficients $u_i(x,t)$ are defined by the projections

$$u_i(x,t) = \langle u(x,t,\xi)\psi_i(\xi)\rangle, \quad i = 0,1,\ldots. \tag{2.8}$$

For notational convenience, we will not distinguish between u and its generalized chaos expansion.

Independent of the choice of basis $\{\psi_i\}_{i=0}^{\infty}$, we can express the mean and variance of $u(x,t,\xi)$ as

$$E(u(x,t,\xi)) = u_0(x,t), \quad \mathrm{Var}(u(x,t,\xi)) = \sum_{i=1}^{\infty} u_i^2(x,t),$$

respectively. Similarly, higher-order statistics, e.g., skewness and kurtosis, can be derived as functions of the gPC coefficients. For practical purposes, (2.7) is truncated to a finite order M, and we set

$$u(x,t,\xi) \approx \sum_{i=0}^{M} u_i(x,t)\psi_i(\xi). \tag{2.9}$$

The number of basis functions $M+1$ is dependent on the number of stochastic dimensions N and the order of truncation of the generalized chaos expansion.

In order to construct a multi-dimensional gPC basis, let $\boldsymbol{\xi} = (\xi_1,\ldots,\xi_N)^T \in \mathbb{R}^N$ be a random vector of input uncertainties defined on the probability space $(\Omega,\mathcal{F},\mathcal{P})$. Assume that the entries of $\boldsymbol{\xi}$ are independent and identically distributed (i.i.d.). For $l = 1,\ldots,d$, let $\{\psi_{k_l}(\xi_l)\}_{k=0}^{\infty}$ be a polynomial basis orthonormal with respect to the measure of the random variable ξ_l. The multi-dimensional gPC basis functions may then be obtained by tensorization of the univariate basis functions $\{\psi_{k_l}(\xi_l)\}_{k=0}^{\infty}$, i.e.,

$$\psi_{\mathbf{k}}(\boldsymbol{\xi}) = \prod_{l=1}^{N} \psi_{k_l}(\xi_l), \tag{2.10}$$

with the multi-index $\boldsymbol{k} \in \mathbb{N}_0^N := \{(k_1, \cdots, k_N) : k_l \in \mathbb{N} \cup \{0\}\}$. In practice, the multi-index $\boldsymbol{k}$ has to be truncated in order to generate a finite cardinality basis. This may be achieved by restricting $\boldsymbol{k}$ to the sets

$$\Lambda_{p,N} := \left\{\boldsymbol{k} \in \mathbb{N}_0^N : \|\boldsymbol{k}\|_1 \leq p\right\} \tag{2.11}$$

or

$$\Gamma_{p,N} := \left\{\boldsymbol{k} \in \mathbb{N}_0^N : k_l \leq p,\ l = 1, \ldots, N\right\} \tag{2.12}$$

to achieve the so-called *complete polynomial* or *tensor polynomial* basis, respectively. The bases defined by the index sets (2.11) and (2.12) are isotropic in the N stochastic dimensions. By replacing p with a dimension-dependent integer p_l, $l = 1, \ldots, N$, anisotropic bases tailored to accuracy requirements for each stochastic dimension may be obtained. For simplicity of notation, we subsequently consider a one-to-one relabeling of the form $\{\psi_k(\boldsymbol{\xi})\}_{k=0}^M$ for the gPC basis $\{\psi_{\boldsymbol{k}}(\boldsymbol{\xi})\}$, $\boldsymbol{k} \in \Lambda_{p,N}$ or $\Gamma_{p,N}$, where $M + 1$ is the cardinality of the gPC basis. In particular, for the complete polynomial basis, the cardinality is given by

$$M + 1 = \frac{(p + N)!}{p!N!},$$

while for the tensor polynomial basis, the cardinality is

$$M + 1 = (p + 1)^N.$$

As an example, consider the case of $p = 5$ and $N = 2$ stochastic dimensions. That means 21 and 36 basis functions for the complete polynomial basis and the tensor polynomial basis, respectively. If we keep $p = 5$ and include 5 stochastic dimensions, $N = 5$, the complete polynomial basis contains 252 basis functions. In contrast, the corresponding tensor polynomial basis contains as many as 7,776 basis functions.

An increase in the number of random parameters corresponds to an exponential increase in the cardinality of the series. This increase quickly leads to infeasible numerical problems and has spurred broad interest in alternative formulations not based on the tensorization introduced earlier. Sparse representations and adaptive techniques [8, 16, 17] are becoming increasingly popular, although their use remains fairly limited for hyperbolic problems. For this reason, and because the fundamental issues related to the numerical treatment of the stochastic Galerkin schemes are well expressed in one-dimensional uncertain problems, we will not discuss this issue further but rather focus on the $N = 1$ case.

The basis $\{\psi_i\}_{i=0}^{\infty}$ is often a set of orthogonal polynomials. Given the two lowest-order polynomials, higher-order polynomials can be generated by the recurrence relation

$$\psi_n(\xi) = (a_n \xi + b_n)\psi_{n-1}(\xi) + c_n \psi_{n-2}(\xi),$$

where the coefficients a_n, b_n, c_n are specific to the class of polynomials.

The truncated chaos series (2.9) may result in solutions that are unphysical. An extreme example is when a strictly positive quantity, say density, with uncertainty within a bounded range is represented by a polynomial expansion with infinite range, for instance Hermite polynomials of standard Gaussian variables. The Hermite series expansion converges to the true density with bounded range in the limit $M \to \infty$, but for a given order of expansion, say $M = 1$, the representation $\rho = \rho_0 + \rho_1 H_1(\xi)$ results in negative density with nonzero probability since the Hermite polynomial H_1 takes arbitrarily large negative values. Similar problems may be encountered also for polynomial representations with bounded support. Polynomial reconstruction of a discontinuity in stochastic space leads to Gibbs oscillations that may yield negative values of an approximation of a solution that is close to zero but strictly positive by definition. Whenever discontinuities are involved, care is needed with the use of global polynomial representations; this caveat underlies most of the development in Chap. 8.

Spectral convergence of the generalized polynomial chaos expansion is observed when the solutions are sufficiently regular and continuous [20], but for general non-linear conservation laws – such as in fluid dynamics problems – the convergence is usually less favorable. Spectral expansion representations are still of interest for these problems because of their potential efficiency with respect to brute force sampling methods and to gain insights from writing the governing equations for the stochastic problem. However, special attention must be devoted to the numerical methodology used. For some problems with steep gradients in the stochastic dimensions, polynomial chaos expansions completely fail to capture the solution [13]. Global methods can still give a superior overall performance, for instance Padé approximation methods based on rational function approximation [4], and hierarchical wavelet methods that are global methods with localized support of each resolution level [11]. These methods do not need input such as mesh refinement parameters, and they are not dependent on the initial discretization of the stochastic space. An alternative to polynomial expansions for non-smooth and oscillatory problems is generalized chaos based on a localization or discretization of the stochastic space [5,15]. Methods based on stochastic discretization such as adaptive stochastic multi-elements [17] and stochastic simplex collocation [19] will be described in some more detail in Sect. 3.2.3. The robust properties of discretized stochastic space can also be obtained by globally defined wavelets, see [11,12]. The next section outlines piecewise linear Haar wavelet chaos, followed by a description of piecewise polynomial multiwavelet generalized chaos. These classes of basis functions are robust to discontinuities.

2.2.2 *Haar Wavelet Expansion*

Haar wavelets are defined hierarchically on different resolution levels, representing successively finer features of the solution with increasing resolution. They have non-overlapping support within each resolution level, and in this sense they are localized. Still, the Haar basis is global due to the overlapping support of wavelets belonging to different resolution levels. Haar wavelets do not exhibit spectral convergence, but avoid the Gibbs phenomenon.

Consider the mother wavelet function defined by

$$\psi^W(y) = \begin{cases} 1 & \text{for } 0 \le y < \frac{1}{2} \\ -1 & \text{for } \frac{1}{2} \le y < 1 \\ 0 & \text{otherwise} \end{cases}. \tag{2.13}$$

Based on (2.13), we get the wavelet family

$$\psi^W_{j,k}(y) = 2^{j/2}\psi^W(2^j y - k), \qquad j = 0, 1, \ldots; \quad k = 0, \ldots, 2^{j-1}.$$

Given the probability measure of the stochastic variable ξ with cumulative distribution function $F_\xi(\xi_0) = \mathscr{P}(\omega : \xi(\omega) \le \xi_0)$, define the basis functions

$$W_{j,k}(\xi) = \psi^W_{j,k}(F_\xi(\xi)).$$

Adding the basis function $W_0(y) = 1$ in $y \in [0, 1]$ and concatenating the indices j and k into $i = 2^j + k$ so that $W_i(\xi) \equiv \psi^W_{n,k}(F_\xi(\xi))$, we can represent any random variable $u(x, t, \xi)$ with finite variance as

$$u(x, t, \xi) = \sum_{i=0}^{\infty} u_i(x, t) W_i(\xi),$$

which is of the form (2.7). Figure 2.1 depicts the first eight basis functions of the generalized Haar wavelet chaos.

2.2.3 Multiwavelet Expansion

The main idea of multiwavelets (MW) is to combine the localized and hierarchical structure of Haar wavelets with the convergence properties of orthogonal polynomials. The procedure of constructing these multiwavelets using Legendre polynomials follows the algorithm in [1] and is outlined in [12]; additional details are included in Appendix A.

Starting with the space $\mathbf{V}_{N_p}$ of polynomials of degree at most N_p defined on the interval $[-1, 1]$, the construction of multiwavelets aims at finding a basis of piecewise polynomials for the orthogonal complement of $\mathbf{V}_{N_p}$ in the space $\mathbf{V}_{N_p+1}$ of polynomials of degree at most $N_p + 1$. Merging the bases of $\mathbf{V}_{N_p}$ and that of the orthogonal complement of $\mathbf{V}_{N_p}$ in $\mathbf{V}_{N_p+1}$, we obtain a piecewise polynomial basis for $\mathbf{V}_{N_p+1}$. Continuing the process of finding orthogonal complements in spaces of increasing degree of piecewise polynomials leads to a basis for $L_2([-1, 1])$.

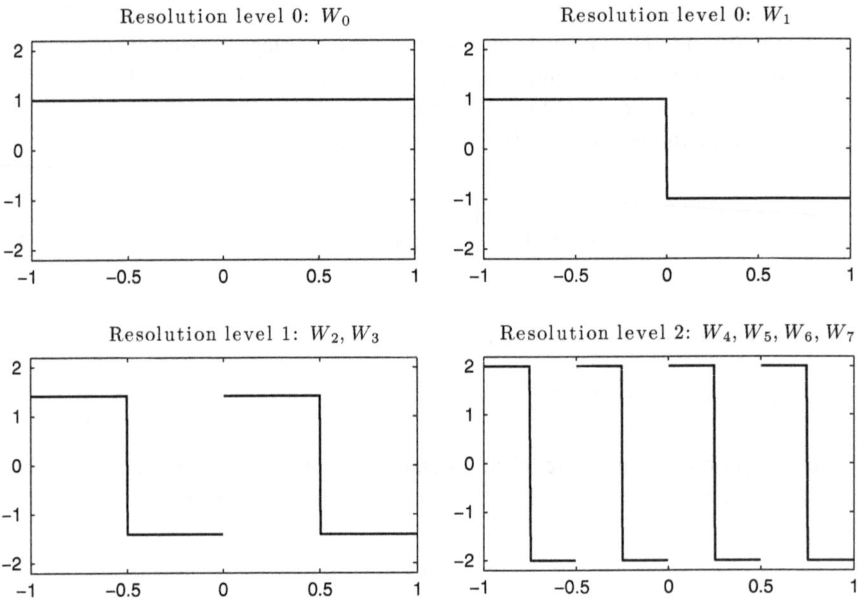

Fig. 2.1 Haar wavelets, resolution levels 0,1,2

We first introduce a smooth polynomial basis on $[-1, 1]$. Let $\{Le_i(\xi)\}_{i=0}^{\infty}$ be the set of Legendre polynomials that are defined on $[-1, 1]$ and orthogonal with respect to the uniform measure. The normalized Legendre polynomials are defined recursively by

$$Le_{j+1}(\xi) = \sqrt{2j+3}\left(\frac{\sqrt{2j+1}}{j+1}\xi Le_j(\xi) - \frac{j}{(j+1)\sqrt{2j-1}}Le_{j-1}(\xi)\right),$$
$$Le_0(\xi) = 1, \quad Le_1(\xi) = \sqrt{3}\xi.$$

The set $\{Le_i(\xi)\}_{i=0}^{N_p}$ is an orthonormal basis for $\mathbf{V}_{N_p}$. Double products are readily computed from (2.6), and higher-order products are precomputed using numerical integration.

Following the algorithm by Alpert [1] (see Appendix A), we construct a set of *mother wavelets* $\{\psi_i^W(\xi)\}_{i=0}^{N_p}$ defined on the domain $\xi \in [-1, 1]$, where

$$\psi_i^W(\xi) = \begin{cases} \pi_i(\xi) & -1 \le \xi < 0 \\ (-1)^{N_p+i+1}\pi_i(\xi) & 0 \le \xi < 1 \\ 0 & \text{otherwise,} \end{cases} \tag{2.14}$$

where $\pi_i(\xi)$ is an ith-order polynomial. By construction, the set of wavelets $\{\psi_i^W(\xi)\}_{i=0}^{N_p}$ are orthogonal to all polynomials of order at most N_p, hence the wavelets are orthogonal to the set $\{Le_i(\xi)\}_{i=0}^{N_p}$ of Legendre polynomials of order at most N_p. Based on translations and dilations of (2.14), we get the wavelet family

$$\psi_{i,j,k}^{W}(\xi)=2^{j/2}\psi_i^{W}(2^j\xi-k), \qquad i=0,\ldots,N_p, \quad j=0,1,\ldots, \quad k=0,\ldots,2^{j-1}.$$

Let $\psi_m(\xi)$ for $m = 0,\ldots,N_p$ be the set of Legendre polynomials up to order N_p, and concatenate the indices i,j,k into $m = (N_p + 1)(2^j + k - 1) + i$ so that $\psi_m(\xi) \equiv \psi_{i,j,k}^{W}(\xi)$ for $m > N_p$. With the MW basis $\{\psi_m(\xi)\}_{m=0}^{\infty}$, we can represent any random variable $u(x,t,\xi)$ with finite variance as

$$u(x,t,\xi) = \sum_{m=0}^{\infty} u_m(x,t)\psi_m(\xi),$$

which is again of the form (2.7). In the computations, we truncate the MW series both in terms of the piecewise polynomial order N_p and the *resolution level* N_r. With the index $j = 0,\ldots,N_r$, we retain $P = (N_p + 1)2^{N_r}$ terms of the MW expansion.

The truncated MW basis is characterized by the piecewise polynomial order N_p and the number of resolution levels N_r, illustrated in Fig. 2.2 for $N_p = 2$ and $N_r = 3$. As special cases of the MW basis, we obtain the Legendre polynomial basis for $N_r = 0$ ($i = j = 0$), and the Haar wavelet basis of piecewise constant functions for $N_p = 0$.

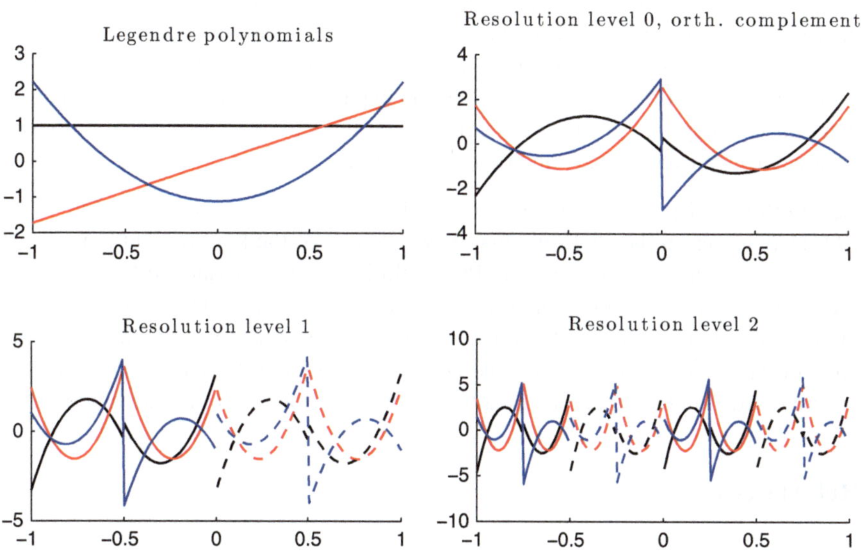

Fig. 2.2 Multiwavelets for $N_p = 2$, $N_r = 3$. Resolution level 0 consists of the first $N_p + 1$ Legendre polynomials and their orthogonal complement. Resolution level $j > 0$ contains $(N_p + 1)2^j$ wavelets each. Each basis function is a piecewise polynomial of order N_p

2.2.4 Choice of Basis Functions for Generalized Chaos

The choice of basis functions for the generalized chaos expansion of a given problem of interest is in general non-trivial. An optimal set of basis functions for the input parameters may be highly inappropriate for the propagation of uncertainty to the output. In particular, this is the case for the nonlinear hyperbolic problems that will be encountered in subsequent chapters. These problems develop discontinuities in finite time, and a polynomial reconstruction will lead to oscillations. The consequence is lack of accuracy or even breakdown of the numerical method.

For smooth problems, the situation is not that severe. Transformations between probability measures allow the use of non-optimal basis functions, e.g., Legendre polynomials to represent normal distributions. The exponential convergence rate of PC expansions is in general not maintained when a non-optimal basis is chosen [21].

2.3 Exercises

2.1. PC formulations of UQ problems typically start from infinite series expansions, ending up with a formulation involving a finite number of PC terms. This truncation introduces a stochastic truncation error that propagates in subsequent operations on the PC series. Verify that the finite order expansion of the product of $F * G$ is different from the product of the expansions of F and G.

2.2. Orthogonal polynomial representations are often used with the hope that a small number of terms are sufficient to accurately represent a given function. Study the truncation error of Hermite expansions of the non-linear functions $\sin(\xi)$, $x^3(\xi)$, $\log(\xi)$, $x^2(\xi)/(3-\xi)$, assuming that ξ is a standard normal random variable. Plot the L_2 error as a function of the order M of the expansion (you need to find functions that can be integrated analytically for the coefficients – or ensure that sufficient accuracy is achieved by the numerical integration).

2.3. Orthogonal polynomials are frequently used to represent PDE solutions in UQ. Depending on the PDE, we may have an idea of the kind of solution we can expect. To accurately represent the PDE solution, it is necessary to know how to accurately represent a function similar to the solution, i.e., how many gPC terms to be retained, and whether the chosen gPC basis is suitable. Consider Legendre polynomial expansion of the sine and Heaviside functions. Consider expansions of different order and compare the resulting approximations with the true function.

References

1. Alpert BK (1993) A class of bases in L_2 for the sparse representations of integral operators. SIAM J Math Anal 24:246–262
2. Askey R, Wilson JA (1985) Some basic hypergeometric orthogonal polynomials that generalize Jacobi polynomials. Memoirs of the American Mathematical Society, vol 319. American Mathematical Society, Providence. http://books.google.com/books?id=9q9o03nD_xsC

3. Cameron RH, Martin WT (1947) The orthogonal development of non-linear functionals in series of Fourier-Hermite functionals. Ann Math 48(2):385–392
4. Chantrasmi T, Doostan A, Iaccarino G (2009) Padé-Legendre approximants for uncertainty analysis with discontinuous response surfaces. J Comput Phys 228:7159–7180. doi:10.1016/j.jcp.2009.06.024, http://dl.acm.org/citation.cfm?id=1595071.1595203
5. Deb MK, Babuška IM, Oden JT (2001) Solution of stochastic partial differential equations using Galerkin finite element techniques. Comput Methods Appl Math 190(48):6359–6372. doi:10.1016/S0045-7825(01)00237-7, http://www.sciencedirect.com/science/article/pii/S0045782501002377
6. Ernst OG, Mugler A, Starkloff HJ, Ullmann E On the convergence of generalized polynomial chaos expansions. DFGSPP 1324(2):317–339 (2010, Preprint). http://citeseerx.ist.psu.edu/viewdoc/summary?doi=10.1.1.178.5189
7. Funaro D (1992) Polynomial approximation of differential equations, 1st edn. Springer, Berlin/New York
8. Gerstner T, Griebel M (1998) Numerical integration using sparse grids. Numer Algorithms 18(3):209–232. http://www.springerlink.com/index/n59w867362x015g2.pdf
9. Ghanem RG, Spanos PD (1991) Stochastic finite elements: a spectral approach. Springer, New York
10. Karhunen K (1946) Zur Spektraltheorie stochastischer Prozesse. Ann Acad Sci Fenn Ser A I Math 34:3–7
11. Le Maître OP, Knio OM, Najm HN, Ghanem RG (2004) Uncertainty propagation using Wiener-Haar expansions. J Comput Phys 197:28–57. doi:10.1016/j.jcp.2003.11.033, http://portal.acm.org/citation.cfm?id=1016237.1016239
12. Le Maître OP, Najm HN, Ghanem RG, Knio OM (2004) Multi-resolution analysis of Wiener-type uncertainty propagation schemes. J Comput Phys 197:502–531. doi:10.1016/j.jcp.2003.12.020, http://portal.acm.org/citation.cfm?id=1017254.1017259
13. Le Maître OP, Najm HN, Pébay PP, Ghanem RG, Knio OM (2007) Multi-resolution-analysis scheme for uncertainty quantification in chemical systems. SIAM J Sci Comput 29:864–889. doi:10.1137/050643118, http://dl.acm.org/citation.cfm?id=1272907.1272926
14. Loève M (1948) Fonctions aleatoires de seconde ordre. In: Levy P (ed) Processus Stochastiques et Mouvement Brownien. Gauthier-Villars, Paris
15. Pettit CL, Beran PS (2006) Spectral and multiresolution Wiener expansions of oscillatory stochastic processes. J Sound Vib 294:752–779
16. Smolyak S (1963) Quadrature and interpolation formulas for tensor products of certain classes of functions. Sov Math Dokl 4:240–243
17. Wan X, Karniadakis GE (2005) An adaptive multi-element generalized polynomial chaos method for stochastic differential equations. J Comput Phys 209:617–642. doi:http://dx.doi.org/10.1016/j.jcp.2005.03.023
18. Wiener N (1938) The homogeneous chaos. Am J Math 60(4):897–936
19. Witteveen JAS, Loeven A, Bijl H (2009) An adaptive stochastic finite elements approach based on Newton-Cotes quadrature in simplex elements. Comput Fluids 38(6):1270–1288. doi:10.1016/j.compfluid.2008.12.002, http://www.sciencedirect.com/science/article/pii/S0045793008002351
20. Xiu D, Karniadakis GE (2002) The Wiener–Askey polynomial chaos for stochastic differential equations. SIAM J Sci Comput 24(2):619–644. doi:http://dx.doi.org/10.1137/S1064827501387826
21. Xiu D, Karniadakis GE (2003) Modeling uncertainty in flow simulations via generalized polynomial chaos. J Comput Phys 187(1):137–167. doi:10.1016/S0021-9991(03)00092-5, http://dx.doi.org/10.1016/S0021-9991(03)00092-5

Chapter 3
Polynomial Chaos Methods

In this chapter we review methods for formulating partial differential equations based on the random field representations outlined in Chap. 2. These include the stochastic Galerkin method, which is the predominant choice in this book, as well as other methods that frequently occur in the literature. We also briefly discuss methods that are not polynomial chaos methods themselves but are viable alternatives.

3.1 Intrusive Methods

In the context of gPC, problem formulations result in a new set of equations that are distinctly different from the original set of equations and thus require the design of new numerical solvers. These solvers are referred to as intrusive methods – as opposed to non-intrusive stochastic methods that exclusively rely on existing deterministic codes.

3.1.1 Stochastic Galerkin Methods

The stochastic Galerkin method was introduced by Ghanem and Spanos in order to solve linear stochastic equations [11]. It relies on a weak problem formulation where the set of solution basis functions (trial functions) is the same as the space of stochastic test functions. Consider a general scalar conservation law defined on a spatial domain Ω_x with boundary Γ_x subject to initial and boundary conditions, given by

$$\frac{\partial u(x,t,\boldsymbol{\xi})}{\partial t} + \frac{\partial f(u(x,t,\boldsymbol{\xi}),\xi)}{\partial x} = 0, \qquad x \in \Omega_x, \quad t \geq 0, \tag{3.1}$$

M.P. Pettersson et al., *Polynomial Chaos Methods for Hyperbolic Partial Differential Equations*, Mathematical Engineering,
DOI 10.1007/978-3-319-10714-1_3

$$\mathscr{L}_\Gamma(u,x,t,\xi) = g(t,\xi), \qquad x \in \Gamma_x, \quad t \geq 0, \tag{3.2}$$

$$u = h(x,\xi), \qquad x \in \Omega_x, \quad t = 0, \tag{3.3}$$

where u is the solution and f is a flux function, for example representing a convection or diffusion process. $\mathscr{L}_\Gamma$ is a boundary condition operator, g is boundary data, and h is the initial function. A weak approximation of 3.1 is obtained by substituting the truncated gPC series of the solution u given by (2.9) into (3.1) and projecting the resulting expression onto the subspace of $L_2(\Omega, \mathscr{P})$ spanned by the (truncated) basis $\{\psi_i(\xi)\}_{i=0}^{M}$. The result is the *stochastic Galerkin formulation* of (3.1),

$$\frac{\partial \boldsymbol{u}_k(x,t)}{\partial t} + \frac{\partial}{\partial x}\left\langle f\left(\sum_{i=0}^{M} u_i\psi_i(\xi),\xi\right),\psi_k\right\rangle = 0, \qquad x \in \Omega_x, \quad t \geq 0, \tag{3.4}$$

$$\langle \mathscr{L}_\Gamma(u,x,t,\xi),\psi_k\rangle = \langle g,\psi_k\rangle\,, \qquad x \in \Gamma_x, \quad t \geq 0, \tag{3.5}$$

$$\langle u,\psi_k\rangle = \langle h,\psi_k\rangle\,, \qquad x \in \Omega_x, \quad t = 0, \tag{3.6}$$

for $k = 0,\ldots,M$, where the inner product $\langle .,. \rangle$ is defined in (2.2).

The problem (3.4)–(3.6) is essentially a deterministic problem in space and time with no explicit dependence on the random variable ξ. Although prevalent in the literature, there are situations, even for linear problems, when it is essential not to restrict the gPC approximations of all input quantities (e.g., material parameters) to the same order M as the gPC representation of the solution. An example is given in Sect. 5.1.2, where we show that the stochastic Galerkin formulation of an advection-diffusion equation leads to an ill-posed problem unless an order at least $2M$ approximation of the diffusion parameter (assuming a single stochastic dimension) is used whenever an order M gPC approximation is used to represent the solution. This is not an argument against stochastic Galerkin methods; it is an argument for numerical analysis. The stochastic Galerkin method has repeatedly been demonstrated to be efficient for a wide range of PDEs and offers a rich framework for analysis.

The stochastic Galerkin formulation (3.4) is an extended deterministic system of coupled equations. In general, it is obviously more complex than the original deterministic problem, and needs to be solved using a tailored numerical scheme. Sometimes, diagonalization of system matrices is possible, resulting in a sequence of simpler problems. In general, however, this is not possible. In later Chapters, we present different strategies to find suitable numerical schemes and elaborate on this topic.

3.1.2 Semi-intrusive Methods

Alternative approaches to generalized polynomial chaos methods have also been presented in the literature. Abgrall et al. [1, 2] developed a semi-intrusive method based on a finite-volume like reconstruction technique in multi-dimensional elements (cells) that span the physical variables and the uncertain parameters. A deterministic problem is obtained by taking conditional expectations, followed by monotonicity-preserving flux reconstructions of an essentially non-oscillatory (ENO) type. This method makes it particularly suitable for non-smooth probability distributions, in contrast to gPC, where the convergence requires the solution to be smooth with respect to the parameters describing the input uncertainty [21]. For further details on the semi-intrusive method, the reader is referred to [1,2].

3.2 Non-intrusive Methods

An alternative to the polynomial chaos approach with stochastic Galerkin projection is to construct empirical probability distributions of the output using multiple samples of solutions corresponding to realizations of the stochastic inputs. Such *non-intrusive methods* do not require modification of existing codes but rely exclusively on repeated runs of the deterministic code, which make them computationally attractive, in particular for complex problems.

3.2.1 Interpolation and Integration Approaches

Stochastic collocation takes a set of solutions $\{u^{(j)}\}$ evaluated at a set $\{\xi^{(j)}\}$ of values of random input ξ and constructs an interpolating polynomial from these solution realizations [3, 15, 25]. A common choice of interpolation polynomials is the set of Lagrange polynomials $\{\mathscr{L}_j^{(M_{int})}(\xi)\}_{j=1}^{M_{int}}$, defined by M_{int} points $\{\xi^{(j)}\}_{j=1}^{M_{int}}$, for which the polynomial interpolant becomes

$$\mathscr{I}u = \sum_{j=1}^{M_{int}} u^{(j)} \mathscr{L}_j(\xi). \tag{3.7}$$

The distribution of the gridpoints $\{\xi^{(j)}\}_{j=1}^{M_{int}}$ is implied by the measure $\mathscr{P}$ of ξ. For instance, we choose $\{\xi^{(j)}\}$ to be the set of Gauss-Legendre quadrature points for the case of uniformly distributed μ, and the set of Gauss-Hermite quadrature points for the case of lognormal μ. The integral statistics of interest, such as moments, may then be approximated by the corresponding quadrature rules. For instance, for some quantity of interest $\langle S(u)\rangle$, we have

$$\langle S(u)\rangle \approx \sum_{j=1}^{M_{int}} S(u^{(j)}) w_j, \tag{3.8}$$

where w_j is the weight corresponding to the quadrature point $\xi^{(j)}$. The quadrature points and weights can be computed through the Golub-Welsch algorithm [12]. Note that there is no need to find the Lagrange polynomials of (3.7) explicitly since $(\mathscr{I}u)(\xi^{(j)}) = u^{(j)}$, and we only need the values of $\mathscr{I}u$ at the quadrature points in (3.8).

This approach is referred to in the literature as a stochastic collocation method and will be used later for comparison with stochastic Galerkin methods. The same numerical integration technique can be applied directly to the evaluation of the polynomial chaos coefficients of u as shown later (this approach is referred to as pseudospectral projection).

Stochastic collocation is similar to other non-intrusive methods such as pseudospectral projection [18] and stochastic point collocation (stochastic response surfaces) [5], in that it relies on evaluating deterministic solutions associated with stochastic quadrature points. The difference is in the postprocessing step where quantities of interest are reconstructed by different means of numerical quadrature. Specifically, in stochastic collocation, quantities of interest are computed directly without representing the solutions as a gPC series. Pseudospectral projection, on the other hand, involves the computation of the polynomial chaos coefficients of u through numerical quadrature. Quantities of interest are then calculated as functions of the polynomial chaos coefficients.

Several investigations of the relative performance of stochastic Galerkin and stochastic collocation methods have been performed (cf. [4, 16, 19]). The significant size of the stochastic Galerkin system may lead to inefficient direct implementations compared to collocation methods and preconditioned iterative Krylov subspace methods. However, the use of suitable techniques for large systems, such as preconditioners, may result in speedup for the solution of stochastic Galerkin systems compared to multiple collocation runs [19]. For high-dimensional problems where the collocation methods tend to become prohibitively expensive, sparse grid-adaptive methods have been suggested to alleviate the computational cost [9].

3.2.2 Spectral Projection

Spectral projection, discrete projection or the pseudospectral approach [18, 23] comprise a set of gPC-based methods relying on deterministic solutions evaluated at sampling points of the parameter domain. These are sometimes referred to as a subgroup of the class of collocation methods [24]. Alternative spectral projection approaches include weighted least squares formulations for determining the gPC coefficients (2.8) [13].

The integrals over the stochastic domain of the gPC projections defined by (2.8) are approximated by sampling or employing numerical quadrature. For multiple stochastic dimensions, sparse grids are attractive, e.g., Smolyak quadrature [14].

The class of non-intrusive polynomial chaos methods also includes methods where the solution is sampled randomly and the statistics are expressed in terms of spectral expansions. The strength of this class of methods is its applicability to situations when the sampling points are not known in parametric form. These methods are alternatively referred to as random discrete L^2 projection, regression, or point collocation [6].

To appreciate the flavor of the methods briefly described in this Section, let $\boldsymbol{\xi}^{(j)}$, $j = 1, \ldots, N$ be a set of realizations of some random vector $\boldsymbol{\xi}$ and let x_k, $k = 1, \ldots, m$ denote spatial discretization points associated with a numerical solver. Then, the truncated gPC approximations based on the numerical PDE solution $\vec{\boldsymbol{u}}_{jk} \approx u(x_k, t, \boldsymbol{\xi}^{(j)})$ for $\boldsymbol{\xi}^{(j)}$ at x_k and time t,

$$u(x_k, t, \boldsymbol{\xi}^{(j)}) \approx \sum_{i=0}^{M} u_i(x_k, t)\psi_i(\boldsymbol{\xi}^{(j)}), \tag{3.9}$$

can be assembled in matrix form,

$$\boldsymbol{U} = \boldsymbol{\psi} \boldsymbol{C}, \tag{3.10}$$

where $\boldsymbol{U} \in \mathbb{R}^{N \times m}$ contains the solution samples $[\boldsymbol{U}]_{jk} = \vec{\boldsymbol{u}}_{jk}$, $\boldsymbol{\psi} \in \mathbb{R}^{N \times M}$ is the matrix of basis function evaluations, and $\boldsymbol{C} \in \mathbb{R}^{M \times m}$ is the matrix of gPC coefficients $[\boldsymbol{C}]_{ji} = \psi_i(\boldsymbol{\xi}^j)$.

The choice of appropriate methods for the solution of (3.10) depends on the size of N and M. For overdetermined problems $N > M$, least squares approaches are applicable and, under certain conditions, yield stable approximations of the coefficient matrix $\boldsymbol{C}$ [17]. For underdetermined systems, (3.10) can be reformulated to the compressive sampling framework [8].

An alternative strategy is to compute the gPC coefficients of $u(x_k, t, \boldsymbol{\xi})$ through direct projection onto one of the basis functions Ψ_m. The result is

$$\left\langle u(x_k, t, \xi^{(j)}), \psi_m \right\rangle = \langle u_m \rangle \tag{3.11}$$

by the orthogonality of the basis. Therefore it is possible to compute the mth coefficient of the gPC expansion simply by integrating the left-hand side using a quadrature method with M_{int} integration points. The choice of M_{int} is not obvious because the integrand at the left-hand side of (3.11) is not a known function and not necessarily a polynomial. This non-intrusive approach is referred to as pseudo-spectral projection.

3.2.3 Stochastic Multi-elements

In multi-element generalized polynomial chaos (ME-gPC), the stochastic domain is decomposed into subdomains, and generalized polynomial chaos is applied element-wise [20, 22]. Local orthogonal polynomial bases can be constructed numerically using the Stieltjes procedure or the modified Chebyshev algorithm [10]. The stochastic Galerkin method may be applied element-wise, and in this sense ME-gPC is an intrusive method.

The multi-element framework allows the combination of refinement of the number of elements (h-refinement) and increase in the order gPC of each element (p-refinement) [20].

3.3 Exercises

3.1. We will consider the problem of computing the gPC coefficients of a given function when analytical expressions are not available. Consider ξ bounded uniformly in $[-1, 1]$. Compute the coefficients of the gPC expansions of ξ^3 and $\sin(\xi)$ for order $M = 1$, $M = 3$, and $M = 5$ using the least-squares approach with different choices of N, i.e., $N = 10$, $N = 100$, $N = 1{,}000$. Select the realization ξ_j as a set of points distributed randomly in the interval $[-1, 1]$.

3.2. Instead of Monte Carlo integration used in Exercise 3.1 for computation of the gPC coefficients, one may choose the points in random space according to a numerical integration rule. For the previous problem use pseudospectral projection with different choices of N, i.e., $N = M$, $N = 1.2M$, $N = 2M$. Select Hermite-Gauss quadrature.

3.3. Compare the results obtained before with the pseudospectral projection with the Clenshaw-Curtis quadrature using the same number of integration points. For the quadrature rule, see Clenshaw and Curtis [7].

3.4. Compute the coefficients of the gPC expansions $1/(1 + 25\xi^2)$ for order $M = 1$, $M = 3$, and $M = 5$ using pseudospectral projection with Hermite-Gauss quadrature and a set of points distributed uniformly in $[-1, 1]$ for the least-squares approach.

References

1. Abgrall R (2008) A simple, flexible and generic deterministic approach to uncertainty quantifications in non linear problems: application to fluid flow problems. Rapport de recherche. http://hal.inria.fr/inria-00325315/en/
2. Abgrall R, Congedo PM, Corre C, Galera S (2010) A simple semi-intrusive method for uncertainty quantification of shocked flows, comparison with a non-intrusive polynomial chaos method. In: ECCOMAS CFD, Lisbon
3. Babuška IM, Nobile F, Tempone R (2007) A stochastic collocation method for elliptic partial differential equations with random input data. SIAM J Numer Anal 45(3):1005–1034
4. Bäck J, Nobile F, Tamellini L, Tempone R (2011) Implementation of optimal Galerkin and collocation approximations of PDEs with random coefficients. ESAIM Proc 33:10–21. doi:10.1051/proc/201133002, http://dx.doi.org/10.1051/proc/201133002

5. Berveiller M, Sudret B, Lemaire M (2006) Stochastic finite element: a non intrusive approach by regression. Eur J Comput Mech 15:81–92
6. Blatman G, Sudret B (2008) Sparse polynomial chaos expansions and adaptive stochastic finite elements using a regression approach. Comptes Rendus Mécanique 336(6):518–523
7. Clenshaw CW, Curtis AR (1960) A method for numerical integration on an automatic computer. Numerische Mathematik 2:197
8. Doostan A, Owhadi H (2011) A non-adapted sparse approximation of PDEs with stochastic inputs. J Comput Phys 230(8):3015–3034. doi:10.1016/j.jcp.2011.01.002, http://dx.doi.org/10.1016/j.jcp.2011.01.002
9. Ganapathysubramanian B, Zabaras N (2007) Sparse grid collocation schemes for stochastic natural convection problems. J Comput Phys 225(1):652–685. doi:10.1016/j.jcp.2006.12.014, http://dx.doi.org/10.1016/j.jcp.2006.12.014
10. Gautschi W (1982) On generating orthogonal polynomials. SIAM J Sci Stat Comput 3:289–317. doi:10.1137/0903018
11. Ghanem RG, Spanos PD (1991) Stochastic finite elements: a spectral approach. Springer, New York
12. Golub GH, Welsch JH (1967) Calculation of Gauss quadrature rules. Tech rep, Stanford
13. Hosder S, Walters R, Balch M (2007) Efficient sampling for non-intrusive polynomial chaos applications with multiple uncertain input variables. In: AIAA-2007-1939, 9th AIAA non-deterministic approaches conference, Honolulu
14. Keese A, Matthies H (2003) Numerical methods and Smolyak quadrature for nonlinear stochastic partial differential equations. Tech rep, Institute of Scientific Computing TU Braunschweig, Brunswick
15. Mathelin L, Hussaini MY (2003) A stochastic collocation algorithm for uncertainty analysis. Tech Rep 2003-212153, NASA Langley Research Center
16. Mathelin L, Hussaini MY, Zang TA, Bataille F (2003) Uncertainty propagation for turbulent, compressible flow in a quasi-1D nozzle using stochastic methods. In: AIAA-2003-4240, 16TH AIAA CFD conference, Orlando, pp 23–26
17. Migliorati G, Nobile F, von Schwerin E, Tempone R (2013) Approximation of quantities of interest in stochastic PDEs by the random discrete L^2 projection on polynomial spaces. SIAM J Sci Comput 35(3):A1440–A1460
18. Reagan MT, Najm HN, Ghanem RG, Knio OM (2003) Uncertainty quantification in reacting-flow simulations through non-intrusive spectral projection. Combust Flame 132(3):545–555
19. Tuminaro RS, Phipps ET, Miller CW, Elman HC (2011) Assessment of collocation and Galerkin approaches to linear diffusion equations with random data. Int J Uncertain Quantif 1(1):19–33
20. Wan X, Karniadakis GE (2005) An adaptive multi-element generalized polynomial chaos method for stochastic differential equations. J Comput Phys 209:617–642. doi:http://dx.doi.org/10.1016/j.jcp.2005.03.023, http://dx.doi.org/10.1016/j.jcp.2005.03.023
21. Wan X, Karniadakis GE (2006) Long-term behavior of polynomial chaos in stochastic flow simulations. Comput Methods Appl Math Eng 195:5582–5596
22. Wan X, Karniadakis GE (2006) Multi-element generalized polynomial chaos for arbitrary probability measures. SIAM J Sci Comput 28(3):901–928. doi:10.1137/050627630, http://dx.doi.org/10.1137/050627630
23. Xiu D (2007) Efficient collocational approach for parametric uncertainty analysis. Commun Comput Phys 2(2):293–309
24. Xiu D (2010) Numerical methods for stochastic computations: a spectral method approach. Princeton University Press, Princeton. http://www.worldcat.org/isbn/9780691142128
25. Xiu D, Hesthaven JS (2005) High-order collocation methods for differential equations with random inputs. SIAM J Sci Comput 27:1118–1139. doi:10.1137/040615201, http://dl.acm.org/citation.cfm?id=1103418.1108714

Chapter 4
Numerical Solution of Hyperbolic Problems

We introduce the spatial discretization schemes for systems of conservation laws that we use later. For smooth problems, summation-by-parts (SBP) operators with weak enforcement of boundary conditions (SAT) are presented. The SBP-SAT schemes allow for the design of stable high-order accurate schemes. Summation by parts is the discrete equivalent of integration by parts and the matrix operators that are presented lead to energy estimates that in turn lead to provable stability. The semidiscrete stability follows naturally from the continuous analysis of well-posedness which provides the boundary conditions in the SBP-SAT technique.

Stability and boundary conditions are the main reason for choosing to use SBP operators. Provable stability means that numerical convergence to the true solution can be guaranteed. There are many alternative numerical schemes that appear to converge, but for the stochastic Galerkin formulations of interest here, we want to be able to prove stability in situations that would otherwise be hard to handle. An example is a solution with multiple discontinuities crossing the numerical boundary. That situation requires stability and correct imposition of boundary conditions.

For non-smooth problems, the need to accurately capture multiple solution discontinuities of hyperbolic stochastic Galerkin systems calls for shock-capturing methods. We outline how the use of the Monotonic Upstream-Centered Scheme for Conservation Laws (MUSCL) with flux limiters and the HLL (after Harten, Lax and van Leer) Riemann solver can be used to treat these cases. We also discuss in brief how to add artificial dissipation and an issue regarding time-integration.

The problems presented here can all be written as one-dimensional conservation laws,

$$\boldsymbol{u}_t + \boldsymbol{f}(\boldsymbol{u})_x = \boldsymbol{0}, \quad x \in D, \quad t \geq 0, \tag{4.1}$$

where $\boldsymbol{u}$ is the solution vector, $\boldsymbol{f}$ is a flux function and D is the spatial domain. When solving (4.1) on a uniform grid, we will use two different classes of numerical

M.P. Pettersson et al., *Polynomial Chaos Methods for Hyperbolic Partial Differential Equations*, Mathematical Engineering,
DOI 10.1007/978-3-319-10714-1_4

schemes. For smooth problems, we use high-order finite difference schemes, and for non-smooth problems, we apply shock-capturing finite volume methods.

SBP operators are used for approximations of spatial derivatives. Their usefulness lies in the possibility of expressing energy decay in terms of known boundary values, exactly as in the continuous case [12, 22]. For smooth problems, one can often prove that the numerical methods are stable and high-order accurate.

Despite the formal high-order accuracy of SBP operators, solutions with multiple discontinuities are not well captured. Instead, a more robust and accurate method for these problems is the MUSCL scheme [30] or the HLL Riemann solver [8] with flux limiting, to be described in Sect. 4.3.

4.1 Summation-by-Parts Operators

In order to obtain stability of the semidiscretized problem for various orders of accuracy and non-periodic boundary conditions, we use discrete operators satisfying a summation-by-parts (SBP) property [9]. Instead of the exact imposition of boundary conditions, we enforce boundary conditions weakly through penalty terms, where the penalty parameters are chosen such that the numerical method becomes stable.

4.1.1 Recipe for Constructing a Scheme

The principles for construction of stable and convergent high-order finite difference schemes for linear and nonlinear boundary conditions are discussed in the context of linear wave propagation problems. The first requirement for obtaining a reliable solution is well-posedness (see [7, 19] as well as Chap. 1 above). A well-posed problem is bounded by the data of the problem and has a unique solution. Uniqueness for linear problems follows more or less directly from the energy estimate. This is, however, not the case for nonlinear problems. Existence is motivated by using a minimal number of boundary conditions. In the rest of this book we assume that existence is not a problem and will not discuss it further. The crucial point in obtaining well-posedness is the boundary conditions. These will be chosen such that an energy estimate is obtained with a minimal number of conditions.

Once we have a well-posed problem, it is meaningful to construct a numerical approximation. We will use high-order finite differences in SBP form and impose the boundary conditions weakly using penalty terms. More details on this productive and well-tested technique are given below. For further reading, see [2, 4, 6, 11, 16–18, 23–26]. A recipe for constructing a stable and convergent scheme when using the SBP-SAT technique is to choose the so-called penalty parameters such that an energy estimate is obtained. For linear problems, this guarantees that the scheme converges to a reliable solution as the mesh size goes to zero. However, as we shall see below, this is not always the case for nonlinear boundary conditions.

4.1.2 *The Continuous Problem*

As a test problem to illustrate the analysis and design of a stable numerical method, we consider a model problem governed by a system of deterministic PDEs; this is directly relevant to the equations deriving from a stochastic Galerkin formulation of a conservation law under uncertainty.

The model problem is given by

$$\boldsymbol{u}_t = \boldsymbol{A}\boldsymbol{u}_x, \quad x \geq 0, \quad \boldsymbol{u} = \begin{pmatrix} v \\ w \end{pmatrix}, \quad \boldsymbol{A} = \begin{pmatrix} 0 & 1 \\ 1 & 0 \end{pmatrix}, \quad \boldsymbol{u}(x,0) = \boldsymbol{u}_0(x), \tag{4.2}$$

which is a linear version of (4.1) with $\boldsymbol{f}(\boldsymbol{u}) = \boldsymbol{A}\boldsymbol{u}$. We have both ingoing and outgoing waves at the boundary $x = 0$, and we will consider both a linear boundary condition $w = \lambda v$ with λ being a constant, and a highly nonlinear boundary condition of the general form $w = F(v)$.

We make the assumption that all solutions decay as x increases, i.e., $\lim_{x\to\infty} u = 0$. This assumption simplifies the analysis and enables us to focus on the interesting boundary $x = 0$. In the rest of this chapter, all boundary terms are evaluated at $x = 0$. The boundary terms for large x are neglected.

4.2 Analysis

Below we outline the standard recipe for constructing a stable scheme for a linear problem. The nonlinear boundary condition will force us to introduce slight modifications.

4.2.1 *Well-Posedness*

The energy method applied to (4.2) yields

$$2\int_0^\infty \boldsymbol{u}^T \boldsymbol{u}_t dx = \|\boldsymbol{u}\|_t^2 = -2vw.$$

To obtain a bounded solution $\|\boldsymbol{u}\|^2 \leq \|\boldsymbol{u}_0\|^2$, the linear and nonlinear boundary conditions must obey

$$\lambda \geq 0, \quad vF(v) \geq 0, \tag{4.3}$$

respectively.

Next we consider uniqueness and start with the linear case. Consider the difference problem for $\Delta \boldsymbol{u} = \boldsymbol{u}_1 - \boldsymbol{u}_2$,

$$\Delta \boldsymbol{u}_t = \boldsymbol{A}\,\Delta \boldsymbol{u}_x, \quad x \geq 0, \quad \Delta \boldsymbol{u} = \begin{pmatrix} \Delta v \\ \Delta w \end{pmatrix}, \quad \Delta \boldsymbol{u}(x,0) = \boldsymbol{0}, \tag{4.4}$$

and the boundary condition $\Delta w = w_1 - w_2 = \lambda \Delta v$. The energy method yields

$$\|\Delta \boldsymbol{u}\|_t^2 = -2\Delta v \Delta w = -\lambda \Delta v^2, \tag{4.5}$$

and clearly the first condition in (4.3) that guarantees a bounded energy also guarantees uniqueness (since we obtain $\|\Delta \boldsymbol{u}\|^2 \leq 0$ by integrating (4.5)). We summarize the result in Theorem 4.1.

Theorem 4.1. *The problem (4.2) with the linear boundary condition $w = \lambda v$ is well-posed in the sense of Definition 1.1 if*

$$\lambda \geq 0. \tag{4.6}$$

In the nonlinear case, we have $\Delta w = w_1 - w_2 = F(v_1) - F(v_2)$. The energy method applied to the difference equation (4.4) yields

$$\|\Delta \boldsymbol{u}\|_t^2 = -2\Delta v \Delta w = -F'(v)\Delta v^2, \tag{4.7}$$

where the intermediate value theorem has been used and $v \in (v_1, v_2)$. Note that an additional condition, namely $F'(v) \geq 0$, must be added onto the second condition in (4.3) which leads to an energy estimate. We summarize the result in Theorem 4.2.

Theorem 4.2. *The problem (4.2) with the nonlinear boundary condition $w = F(v)$ is well-posed in the sense of Definition 1.1 if*

$$vF(v) \geq 0 \quad \text{and} \quad F'(v) \geq 0. \tag{4.8}$$

4.2.2 Stability

We use high-order finite difference techniques in SBP form and impose the boundary conditions weakly using the simultaneous approximation term (SAT) technique. The first and second derivative SBP operators were introduced in [9, 22] and [4, 11], respectively. The discretized solution $\vec{\boldsymbol{u}}$ is represented as a grid function defined in Sect. 1.1.2. For the first derivative, we use the discrete approximation $\boldsymbol{u}_x \approx \boldsymbol{P}^{-1}\boldsymbol{Q}\vec{\boldsymbol{u}}$, where subscript x denotes a partial derivative with respect to x and $\boldsymbol{Q}$ satisfies

$$\boldsymbol{Q} + \boldsymbol{Q}^T = \text{diag}(-1, 0, \ldots, 0, 1) \equiv \tilde{\boldsymbol{B}}. \tag{4.9}$$

Additionally, $\boldsymbol{P}$ must be symmetric and positive definite in order to define a discrete norm. Operators of order $2n$, $n \in \mathbb{N}$, in the interior of the domain, are combined with boundary closures of order n. It is possible to design operators with higher-order accuracy at the boundary, but this would require $\boldsymbol{P}$ to have nonzero off-diagonal entries. We restrict ourselves to diagonal matrices $\boldsymbol{P}$ since the proofs of stability to be presented in later sections rely on this assumption.

For the approximation of the second derivative, we can either use the first derivative operator twice, or use $\vec{\boldsymbol{u}}_{xx} \approx \boldsymbol{P}^{-1}(-\boldsymbol{M} + \tilde{\boldsymbol{B}}\boldsymbol{D})\vec{\boldsymbol{u}}$, where $\boldsymbol{M} + \boldsymbol{M}^T \geq \boldsymbol{0}$, $\tilde{\boldsymbol{B}}$ is given by (4.9), and $\boldsymbol{D}$ is a first-derivative approximation at the boundaries, i.e.,

$$\boldsymbol{D} = \frac{1}{\Delta x}\begin{bmatrix} d_1 & d_2 & d_3 & \dots & & \\ & 1 & & & & \\ & & \ddots & & & \\ & & & 1 & & \\ & \dots & -d_3 & -d_2 & -d_1 & \end{bmatrix}, \tag{4.10}$$

where d_i, $i = 1, 2, 3, \ldots$, are scalar values leading to a consistent first-derivative approximation at the boundaries.

The semidiscrete formulation of (4.2) with the *weakly* enforced boundary condition is

$$\vec{\boldsymbol{u}}_t = \left(\boldsymbol{P}^{-1}\boldsymbol{Q} \otimes \boldsymbol{A}\right)\vec{\boldsymbol{u}} + \boldsymbol{P}^{-1}\vec{\boldsymbol{e}}_0 \otimes \boldsymbol{\Sigma}\boldsymbol{B}\left(\vec{\boldsymbol{u}}_0\right), \tag{4.11}$$

where $\vec{\boldsymbol{e}}_0 = (1, 0, \cdots, 0)^T$, $\otimes$ is the Kronecker product, $\boldsymbol{\Sigma} = (\boldsymbol{\Sigma}_1, \boldsymbol{\Sigma}_2)$ is the penalty matrix, $\vec{\boldsymbol{u}} = \left(\vec{\boldsymbol{u}}_0, \vec{\boldsymbol{u}}_1, \cdots, \vec{\boldsymbol{u}}_m\right)^T$, $\vec{\boldsymbol{u}}_i = (v_i, w_i)^T$ and

$$\vec{B}_s(\vec{\boldsymbol{u}}_0) = (1, 1)^T\left[w_0 - F(v_0)\right]. \tag{4.12}$$

We augment (4.11),(4.12) with the initial condition $\vec{\boldsymbol{u}}(t = 0) = \vec{\boldsymbol{u}}_0$.

Note that we have expressed both the linear and nonlinear standard boundary condition in the same functional form ($w_0 = F(v_0)$). We have used a SBP difference operator $\boldsymbol{P}^{-1}\boldsymbol{Q}$ (see [9,22]) and imposed the boundary conditions weakly using the SAT technique [3]. The SBP difference operators satisfy the relation (4.9) and hence they mimic integration by parts perfectly. More details on the weak imposition of boundary and interface conditions using the SAT technique will be given below, and further details can be found in [4, 11, 16–18, 23–25].

Multiplying (4.11) from the left with $\vec{\boldsymbol{u}}^T(\boldsymbol{P} \otimes \boldsymbol{I})$ and the choice $\boldsymbol{\Sigma}_1 = 1$ and $\boldsymbol{\Sigma}_2 = 0$ leads to

$$\frac{d}{dt}\|\vec{\boldsymbol{u}}\|_h^2 = -2v_0 F(v_0), \tag{4.13}$$

which is completely similar to the continuous estimates in both the linear and nonlinear case. We summarize the result below.

Theorem 4.3. *The approximation (4.11) of the problem (4.2) is stable in the sense of Definition 1.3 for both the linear (4.6) and nonlinear (4.8) boundary condition if the penalty coefficients $\Sigma_1 = 1$ and $\Sigma_2 = 0$ are used.*

Note that the conditions (4.6) and (4.8) that lead to well-posedness in the continuous case are necessary for stability.

4.2.3 Convergence for Finite Time

We will derive the error equation and investigate under which requirements the numerical solution converges to the analytic solution using weak non-characteristic boundary conditions.

By inserting the analytical solution $\vec{\boldsymbol{u}}_{exact}$ (projected onto the mesh) in (4.11) and subtracting (4.11) we obtain the error equation

$$\vec{e}_t = \left(\boldsymbol{P}^{-1}\boldsymbol{Q} \otimes \boldsymbol{A}\right)\vec{e} + \boldsymbol{P}^{-1}e_0 \otimes \boldsymbol{\Sigma}\boldsymbol{B}\left(\vec{e}_0\right) + \vec{t}_e, \tag{4.14}$$

where $\vec{\boldsymbol{e}} = \vec{\boldsymbol{u}}_{exact} - \vec{\boldsymbol{u}}, \vec{\boldsymbol{e}} = (e_0, e_1, \cdots, e_m)^T, \vec{\boldsymbol{e}}_i = (\Delta v_i, \Delta w_i)^T$ is the error in the numerical solution, $\vec{\boldsymbol{t}}_e = \vec{O}(\Delta x^p)$ is the truncation error and

$$\vec{B}_s(\vec{\boldsymbol{e}}_0) = (1,1)^T\left[\Delta w_0 - (F(\bar{v}_0) - F(v_0))\right]. \tag{4.15}$$

The initial data is zero (we initiate the numerical solution with the exact initial data projected onto the grid), i.e., $\vec{\boldsymbol{e}}(0) = \boldsymbol{0}$.

We assume that the truncation error $\vec{\boldsymbol{t}}_e = \vec{O}(\Delta x^p)$ is uniform in accuracy, although, in reality, the accuracy close to the boundaries is lower (see [22]). This is especially true for the diagonal norm $\boldsymbol{P}$ which is needed in many cases for stability (see for example [14, 17] and examples in subsequent Chapters of this book).

By multiplying (4.14) from the left with $\vec{\boldsymbol{e}}^T(\boldsymbol{P} \otimes \boldsymbol{I})$, we obtain

$$\frac{d}{dt}\|\vec{\boldsymbol{e}}\|_h^2 = -2\Delta v_0(F(\bar{v}_0) - F(v_0)) + 2\vec{\boldsymbol{e}}^T(\boldsymbol{P} \otimes \boldsymbol{I})\vec{\boldsymbol{t}}_e, \tag{4.16}$$

where $\Sigma_1 = 1$ and $\Sigma_2 = 0$ have been used. In the linear case, the first term is negative by the fact that condition (4.6) holds. In the nonlinear case, the intermediate value theorem in combination with the second condition in (4.8) leads to the same result.

The negative contribution of the first term in (4.16) and the standard inequality

$$2(u,v) \leq \eta\|u\|^2 + (1/\eta)\|v\|^2 \tag{4.17}$$

leads to

$$\frac{d}{dt}\|\vec{e}\|_h^2 \leq \eta\|\vec{e}\|_h^2 + (1/\eta)\|\vec{t}_e\|_h^2. \tag{4.18}$$

Time integration of (4.18) leads to the final accuracy result

$$\|\vec{e}(T)\|_h^2 \leq \frac{e^{\eta T}}{\eta}\int_0^T e^{-\eta t}\|\vec{t}_e\|_h^2 dt = \vec{O}(\Delta x^{2p}), \tag{4.19}$$

which we summarize below.

Theorem 4.4. *The solution of the approximation (4.11) converges to the solution of the problem (4.2) with the linear (4.6) and nonlinear (4.8) boundary conditions if the penalty coefficients $\Sigma_1 = 1$ and $\Sigma_2 = 0$ are used.*

4.2.4 An Error Bound in Time

As a final exercise, we will show that under reasonable assumptions, the error growth in time is bounded even for long times, see [1] and in particular [15]. Equation (4.16) can be written as

$$\frac{d}{dt}\|\vec{e}\|_h^2 \leq -2C_0|\vec{e}_0|^2 + 2\|\vec{e}\|\|\vec{t}_e\|, \tag{4.20}$$

where C_0 is an appropriate non-zero constant. By expanding the left-hand side as $\frac{d}{dt}\|\vec{e}\|_h^2 = 2\|\vec{e}\|_h\frac{d}{dt}\|\vec{e}\|_h$, we get

$$\frac{d}{dt}\|\vec{e}\|_h \leq -\eta(t)\|\vec{e}\|_h + \|\vec{t}_e\|, \tag{4.21}$$

where $\eta(t) = C_0|\vec{e}_0|^2/\|\vec{e}\|_h^2$.

For the sake of argument, we assume that $\eta(t) = \eta = const.$ independent of time. In that case, we can integrate (4.21) and obtain

$$\|\vec{e}(T)\|_h \leq e^{-\eta T}\int_0^T e^{\eta t}\|\vec{t}_e(t)\|_h dt.$$

The estimate

$$\|\vec{t}_e(t)\|_h \leq \max_{0\leq t\leq T}\|\vec{t}_e(t)\|_h = (\|\vec{t}_e\|_h)_{max}$$

leads to the final error bound

$$\|\vec{e}(T)\|_h \leq (\|\vec{t}_e\|_h)_{max} \frac{(1 - e^{-\eta T})}{\eta}. \tag{4.22}$$

In the case of a time-dependent $\eta(t)$, not much is changed as long as $\eta(t)$ is non-negative and monotonically increasing. The conclusion (4.22) still holds (see [15]). The weakly imposed boundary conditions lead to an error bound in time.

4.2.5 Artificial Dissipation Operators

An artificial dissipation operator is a discretized even-order derivative which is added to the system to allow stable and accurate solutions to be obtained in the presence of solution discontinuities. The artificial dissipation is designed to transform the global discretization into a one-sided operator close to the shock location. Depending on the accuracy of the difference scheme, this transformation requires one or more dissipation operators. All dissipation operators used here are of the form

$$\boldsymbol{A}_{2k} = -\Delta x \boldsymbol{P}^{-1} \tilde{\boldsymbol{D}}_k^T \boldsymbol{B}_w \tilde{\boldsymbol{D}}_k, \tag{4.23}$$

where $\boldsymbol{P}^{-1}$ is the diagonal norm of the first derivative as before, $\tilde{\boldsymbol{D}}$ is an approximation of $(\Delta x)^k \partial^k / \partial x^k$, and $\boldsymbol{B}_w$ is a diagonal positive definite matrix. In most cases here, $\boldsymbol{B}_w$ is replaced by a single constant β_w. An appropriate choice of dissipation constant results in an upwind scheme, suitable for problems where shocks evolve. For further reading about the design of artificial dissipation operators we refer to [12]. Here we focus on shock-capturing schemes, such as the MUSCL scheme in the next section.

4.3 Shock-Capturing Methods

For finite volume methods on structured grids, we partition the computational domain into cells of equal size Δx. Solution values $\vec{\boldsymbol{u}}_j$ are defined as cell averages of cell j, and fluxes are defined on the edges of the cells.

4.3.1 MUSCL Scheme

MUSCL (Monotone Upstream-centered Schemes for Conservation Laws) is a finite volume method suitable for conservation laws with discontinuous solutions that was introduced in [30]. Let m be the number of spatial grid cells and $\Delta x = 1/m$, and let $\vec{\boldsymbol{u}} = (\boldsymbol{u}_1^T, \boldsymbol{u}_2^T, \ldots, \boldsymbol{u}_m^T)^T$ be the spatial discretization of $\boldsymbol{u} = (u_1, \ldots, u_n)^T$. The semidiscretized form of (4.1) is given by

$$\frac{d\boldsymbol{u}_j}{dt} + \frac{\boldsymbol{f}_{j+1/2} - \boldsymbol{f}_{j-1/2}}{\Delta x} = 0, \quad j = 1, \ldots, m. \tag{4.24}$$

We define the flux function $\boldsymbol{f}_{j+1/2}$ at the interface between the cells j and $j+1$ as defined by the two states $\boldsymbol{u}^L_{j+\frac{1}{2}}$ and $\boldsymbol{u}^R_{j+\frac{1}{2}}$, obtained with flux limiters. For the MUSCL scheme, we use the numerical flux function

$$\boldsymbol{f}_{j+\frac{1}{2}} = \frac{1}{2}\left(\boldsymbol{f}(\boldsymbol{u}^L_{j+\frac{1}{2}}) + \boldsymbol{f}(\boldsymbol{u}^R_{j+\frac{1}{2}})\right) + \frac{1}{2}|\tilde{\boldsymbol{J}}_{j+\frac{1}{2}}|\left(\boldsymbol{u}^L_{j+\frac{1}{2}} - \boldsymbol{u}^R_{j+\frac{1}{2}}\right), \tag{4.25}$$

where $\tilde{\boldsymbol{J}}$ is an approximation of the flux Jacobian $\boldsymbol{J} = \partial \boldsymbol{f}/\partial \boldsymbol{u}$, from which is derived the absolute value $|\tilde{\boldsymbol{J}}_{j+\frac{1}{2}}|$ given by

$$|\tilde{\boldsymbol{J}}_{j+\frac{1}{2}}| = \boldsymbol{X}\left|\boldsymbol{\Lambda}(\boldsymbol{u}_{j+\frac{1}{2}})\right|\boldsymbol{X}^{-1} = \frac{1}{2}\boldsymbol{X}\left|\boldsymbol{\Lambda}(\boldsymbol{u}^L_{j+\frac{1}{2}}) + \boldsymbol{\Lambda}(\boldsymbol{u}^R_{j+\frac{1}{2}})\right|\boldsymbol{X}^{-1}, \tag{4.26}$$

where $\boldsymbol{\Lambda}$ is a diagonal matrix with the eigenvalues of $\tilde{\boldsymbol{J}}$ and $\boldsymbol{X}$ is the eigenvector matrix. $\tilde{\boldsymbol{J}}$ can be an average of the true Jacobian evaluated at the discretization points or a Roe average matrix [20].

The left and right solution states are given by

$$\boldsymbol{u}^L_{j+\frac{1}{2}} = \boldsymbol{u}_j + 0.5\boldsymbol{\phi}(\boldsymbol{r}_j)(\boldsymbol{u}_{j+1} - \boldsymbol{u}_j) \quad \text{and} \quad \boldsymbol{u}^R_{j+\frac{1}{2}} = \boldsymbol{u}_{j+1} - 0.5\boldsymbol{\phi}(\boldsymbol{r}_{j+1})(\boldsymbol{u}_{j+2} - \boldsymbol{u}_{j+1}),$$

respectively. The vector-vector products are assumed entry-wise above. The flux limiter $\boldsymbol{\phi}(\boldsymbol{r}_j)$ takes the argument $\boldsymbol{r}_j = (r_{1,j}, \ldots, r_{n,j})$ with $r_{i,j} = (u_{i,j} - u_{i,j-1})/(u_{i,j+1} - u_{i,j})$ for $i = 1, \ldots, n$. As a special case, $\boldsymbol{\phi} = \boldsymbol{0}$ results in the first-order accurate upwind scheme. Second-order accurate and total variation diminishing schemes are obtained for ϕ that are restricted to the region

$$\begin{aligned}
&\phi(r) = 0, \quad r \le 0,\\
&r \le \phi(r) \le 2r, \quad 0 \le r \le 1,\\
&1 \le \phi(r) \le r, \quad 1 \le r \le 2,\\
&1 \le \phi(r) \le 2, \quad r \ge 2,\\
&\phi(1) = 1,
\end{aligned}$$

as defined in [27]. The minmod, van Leer and superbee limiters that are used here are all second order and total variation diminishing. For a more detailed description of the MUSCL scheme (see e.g., [10]).

4.3.2 HLL Riemann Solver

As a simpler alternative to the MUSCL-Roe solver, we use the HLL (after Harten, Lax and van Leer) Riemann solver introduced in [8] and further developed in [5]. Instead of computing the Roe average matrix needed for the Roe fluxes (4.25) and (8.9), only the fastest signal velocities need be estimated for the HLL solver. These signal velocities S_L and S_R are the estimated maximum and minimum eigenvalues of the flux Jacobian $\boldsymbol{J} = \partial \boldsymbol{f}/\partial \boldsymbol{u}$. Note that this simplification is particularly important when we derive systems of PDEs from the stochastic Galerkin formulation, in which it is computationally expensive to analytically determine the Roe matrix.

At the interface between the cells j and $j+1$, the HLL flux is defined by

$$\boldsymbol{f}_{j+\frac{1}{2}} = \begin{cases} \boldsymbol{f}\left(\boldsymbol{u}^L_{j+\frac{1}{2}}\right) & \text{if } S_L \geq 0 \\ \frac{S_R \boldsymbol{f}\left(\boldsymbol{u}^L_{j+\frac{1}{2}}\right) - S_L \boldsymbol{f}\left(\boldsymbol{u}^R_{j+\frac{1}{2}}\right) + S_L S_R \left(\boldsymbol{u}^R_{j+\frac{1}{2}} - \boldsymbol{u}^L_{j+\frac{1}{2}}\right)}{S_R - S_L} & \text{if } S_L < 0 < S_R \\ \boldsymbol{f}\left(\boldsymbol{u}^R_{j+\frac{1}{2}}\right) & \text{if } S_R \leq 0 \end{cases}. \tag{4.27}$$

In general, obtaining accurate eigenvalue estimates may be computationally costly. However, for certain choices of stochastic basis functions in combination with known eigenvalues of the deterministic system, we derive analytical expressions for the stochastic Galerkin system eigenvalues (cf. Chap. 8 and Appendix B).

The HLL flux approximates the solution by assuming three states separated by two waves. In the deterministic case, this approximation is known to fail in capturing contact discontinuities and material interfaces of solutions to systems with more than two waves [28]. For the Euler equations, the contact surface can be restored by using the HLLC (Harten-Lax-van Leer-Contact) solver where three waves are assumed [29]. The stochastic Galerkin system is a multiwave generalization of the deterministic case, and similar problems in capturing missing waves are expected. However, the robustness and simplicity of the HLL solver makes it a potentially more suitable choice compared to other Riemann solvers that are theoretically more accurate, but also more sensitive to ill-conditioning of the system matrix.

4.4 Time Integration

Since the stability analysis is based on semidiscretization in space, with time left continuous, we focus more on the spatial discretization than on the time integration. However, the choice of time integration procedure is indeed important and affects the stability properties. In this section we will briefly outline a few important considerations for time integration.

All numerical results presented in this book are based on *explicit* time integration methods, where the new solution is updated directly from operations applied to the previous solution only. For the problems presented, it is also possible to use *implicit* time integration methods, where one solves a system of equations involving the new solution and the previous solution. In particular, the steady advection-diffusion problem of Chap. 5 could benefit from an implicit time integration method, and we encourage the interested reader to try this by rewriting the Matlab scripts that come with Chap. 5. For a more detailed exposition on the basics on implicit and explicit methods and some of their features, we refer to numerical analysis textbooks, e.g., [13].

We will limit our comments on explicit methods to the additional effects that occur for stochastic problems. Standard explicit time integration methods, e.g., forward Euler and Runge-Kutta methods, are conditionally stable for hyperbolic problems. The stability region is determined by the eigenvalues in the complex plane of the total semidiscretized system matrix. This results in a time-step restriction. The maximum time-step for the equations resulting from the stochastic Galerkin formulation is typically more severe than that of the corresponding deterministic problem, but for moderate variance, the difference is not significant. To understand why this is the case, consider a time-dependent random scalar ODE,

$$\frac{du}{dt} = \xi u,$$

where ξ is a known real-valued random variable and negative almost surely. The corresponding stochastic Galerkin problem of order M is

$$\frac{d\boldsymbol{u}}{dt} = \boldsymbol{A}\boldsymbol{u}, \quad \boldsymbol{u} = (u_0, \dots, u_M)^T,$$

where $[\boldsymbol{A}]_{ij} = \langle \xi \psi_i \psi_j \rangle$. A forward Euler discretization of the stochastic Galerkin system yields

$$\vec{\boldsymbol{u}}^{n+1} = (\boldsymbol{I} + \Delta t \boldsymbol{A})\vec{\boldsymbol{u}}^n,$$

where n is the time index and Δt is the time-step. For a fixed real value, say the expectation, $\xi = \bar{\xi}$, the forward Euler discretization of the corresponding deterministic ODE is

$$\vec{u}^{n+1} = (1 + \Delta t \bar{\xi})\vec{u}^n.$$

For stability, we require $\Delta t \leq 2/|\bar{\xi}|$ for the deterministic problem, and $\Delta t \leq 2/(\max_i |\lambda_i|)$ for the stochastic Galerkin problem. Here λ_i, $i = 0, \dots, M$ are the eigenvalues of $\boldsymbol{A}$. For the random variables and associated orthonormal polynomials of interest here, the eigenvalues of $\boldsymbol{A}$ will be spread around the mean $\bar{\xi}$ and the spectral radius of $\boldsymbol{A}$ will increase with the variance in ξ, cf. [21]. Thus, the time-step restriction will be determined by the extreme eigenvalues. Note that for the example provided, an increase in variance for a given probability distribution leads to a stronger time-step restriction, but it is possible to construct examples including large variance without stronger time-step restriction compared to the deterministic case. The point is that variability *typically* leads to restrictions since we try to solve for an entire range of a random space simultaneously, as opposed to solving a problem for a single parameter value.

4.5 Exercises

In the exercises of this Chapter, we will study three sets of PDEs. As the first task, consider the coupling of the two scalar advection equations

$$\begin{aligned} u_t + au_x &= 0, \ -1 \leq x \leq 0 \\ v_t + av_x &= 0, \ 0 \leq x \leq 1 \\ u(0,t) &= v(0,t). \end{aligned} \tag{4.28}$$

As the second task, consider the scalar advection-diffusion problem,

$$\begin{aligned} u_t + au_x &= (\epsilon u_x)_x, \quad 0 \leq x \leq 1 \\ u(0,t) &= g_0(t), \\ u_x(1,t) &= g_1(t), \\ u(x,0) &= f(x), \quad 0 \leq x \leq 1. \end{aligned} \tag{4.29}$$

Both a and ϵ are positive, a is constant and ϵ varies in space and time. As the third task, consider the scalar wave propagation problem,

$$\begin{aligned} u_t + au_x + bu_y &= 0, \quad (x,y) \in \Omega, \\ Lu &= g(x,y,t), \quad (x,y) \in \delta\Omega \\ u(x,y,0) &= f(x,y), \quad (x,y) \in \Omega. \end{aligned} \tag{4.30}$$

The wave propagation direction $\bar{a} = (a,b)$ is constant, and both a and b are positive.

4.1. Introduce a mesh and write up the semidiscrete formulation of problem (4.28) using SBP operators and the SAT-penalty formulation for the boundary and interface conditions. Repeat for different operators and meshes on the domains.

4.2. Prove stability of the semidiscrete formulation for (4.28) using the energy method (determine the penalty parameters). This means that both the left boundary treatment and the interface must be stable.

4.3. Use the energy method on (4.29) and show that the boundary conditions with zero data lead to a well-posed problem with an energy estimate.

4.4. Discretize (4.29) using the SBP-SAT technique. Construct penalty terms for the boundary conditions. For the boundary condition at zero, use a penalty term of the form $\sigma_0 \boldsymbol{P}^{-1}\boldsymbol{D}^T(\vec{u}_0 - 0)\vec{e}_0$. Prove stability and show that the resulting semi-discrete energy estimate with zero boundary data is similar to the continuous one derived in 4.3 above.

4.5. Replace the boundary condition at zero in (4.29) with $au - \epsilon u_x = g_0(t)$. Repeat the same tasks as in 4.3 and 4.4 (replace the penalty term at zero with a new one) above but now with nonzero data.

4.6. Let $\Omega = [0,1] \times [0,1]$ be the unit square. Use the energy method on (4.30) to determine the boundary operator L and where to impose boundary conditions.

4.7. Discretize (4.30) using high-order finite difference methods (FDM) on SBP form and use penalty terms for the boundary condition. The approximation will look like

$$\begin{aligned}\vec{u}_t &+ a(\boldsymbol{P}_x^{-1}\boldsymbol{Q}_x \otimes \boldsymbol{I}_y)\vec{u} + b(\boldsymbol{I}_x \otimes \boldsymbol{P}_y^{-1}\boldsymbol{Q}_y)\vec{u} \\ &= (\boldsymbol{P}_x^{-1} \otimes \boldsymbol{P}_y^{-1})((\boldsymbol{E}_0 \otimes \boldsymbol{\Sigma}_x) + (\boldsymbol{\Sigma}_y \otimes \boldsymbol{E}_0))(\vec{u} - \vec{g}).\end{aligned}$$

$(\boldsymbol{E}_0)_{11}$ is one, the rest of $(\boldsymbol{E}_0)_{ij}$ is zero. Use the energy method and determine $\boldsymbol{\Sigma}_x$ and $\boldsymbol{\Sigma}_y$ so that the approximation is stable. Assume that $\boldsymbol{P}_x$ and $\boldsymbol{P}_y$ are diagonal.

References

1. Abarbanel S, Ditkowski A, Gustafsson B (2000) On error bounds of finite difference approximations to partial differential equations – temporal behavior and rate of convergence. J Sci Comput 15(1):79–116
2. Berg J, Nordström J (2011) Stable Robin solid wall boundary conditions for the Navier-Stokes equations. J Comput Phys 230:7519–7532
3. Carpenter MH, Gottlieb D, Abarbanel S (1994) Time-stable boundary conditions for finite-difference schemes solving hyperbolic systems: methodology and application to high-order compact schemes. J Comput Phys 111(2):220–236. doi:http://dx.doi.org/10.1006/jcph.1994.1057
4. Carpenter MH, Nordström J, Gottlieb D (1999) A stable and conservative interface treatment of arbitrary spatial accuracy. J Comput Phys 148(2):341–365. doi:http://dx.doi.org/10.1006/jcph.1998.6114
5. Einfeld B (1988) On Godunov-type methods for gas dynamics. SIAM J Numer Anal 25(2):294–318. doi:10.1137/0725021, http://dx.doi.org/10.1137/0725021
6. Gong J, Nordström J (2011) Interface procedures for finite difference approximations of the advection-diffusion equation. J Comput Appl Math 236(5):602–620
7. Gustafsson B, Kreiss HO, Oliger J (1995) Time dependent problems and difference methods. Wiley, New York
8. Harten A, Lax PD, van Leer B (1983) On upstream differencing and Godunov-type schemes for hyperbolic conservation laws. SIAM Rev 25(1):35–61. http://www.jstor.org/stable/2030019

9. Kreiss HO, Scherer G (1974) Finite element and finite difference methods for hyperbolic partial differential equations. In: Mathematical aspects of finite elements in partial differential equations. Academic, New York, pp 179–183
10. LeVeque RJ (2002) Finite volume methods for hyperbolic problems. Cambridge University Press, Cambridge
11. Mattsson K, Nordström J (2004) Summation by parts operators for finite difference approximations of second derivatives. J Comput Phys 199(2):503–540. doi:10.1016/j.jcp.2004.03.001, http://dx.doi.org/10.1016/j.jcp.2004.03.001
12. Mattsson K, Svärd M, Nordström J (2004) Stable and accurate artificial dissipation. J Sci Comput 21(1):57–79
13. Moin P (2010) Fundamentals of engineering numerical analysis. Cambridge University Press, New York. http://books.google.no/books?id=uvpwKK7ZVwMC
14. Nordström J (2006) Conservative finite difference formulations, variable coefficients, energy estimates and artificial dissipation. J Sci Comput 29(3):375–404. doi:http://dx.doi.org/10.1007/s10915-005-9013-4
15. Nordström J (2007) Error bounded schemes for time-dependent hyperbolic problems. SIAM J Sci Comput 30(1):46–59. doi:10.1137/060654943
16. Nordström J, Carpenter MH (1999) Boundary and interface conditions for high-order finite-difference methods applied to the Euler and Navier-Stokes equations. J Comput Phys 148(2):621–645. doi:http://dx.doi.org/10.1006/jcph.1998.6133
17. Nordström J, Carpenter MH (2001) High-order finite difference methods, multidimensional linear problems, and curvilinear coordinates. J Comput Phys 173(1):149–174. doi:http://dx.doi.org/10.1006/jcph.2001.6864
18. Nordström J, Gong J, van der Weide E, Svärd M (2009) A stable and conservative high order multi-block method for the compressible Navier-Stokes equations. J Comput Phys 228(24):9020–9035
19. Nordström J, Svärd M (2005) Well-posed boundary conditions for the Navier-Stokes equations. SIAM J Numer Anal 43(3):1231–1255
20. Roe PL (1981) Approximate Riemann solvers, parameter vectors, and difference schemes. J Comput Phys 43(2):357–372. doi:10.1016/0021-9991(81)90128-5, http://www.sciencedirect.com/science/article/B6WHY-4DD1MT3-6G/2/d95f5f5f3b2f002fe5d1fee93f0c6cf8
21. Sonday BE, Berry RD, Najm HN, Debusschere BJ (2011) Eigenvalues of the Jacobian of a Galerkin-projected uncertain ODE system. SIAM J Sci Comput 33:1212–1233. doi:http://dx.doi.org/10.1137/100785922
22. Strand B (1994) Summation by parts for finite difference approximations for d/dx. J Comput Phys 110(1):47–67. doi:http://dx.doi.org/10.1006/jcph.1994.1005
23. Svärd M, Carpenter M, Nordström J (2007) A stable high-order finite difference scheme for the compressible Navier-Stokes equations: far-field boundary conditions. J Comput Phys 225(1):1020–1038
24. Svärd M, Nordström J (2006) On the order of accuracy for difference approximations of initial-boundary value problems. J Comput Phys 218(1):333–352. doi:10.1016/j.jcp.2006.02.014, http://dx.doi.org/10.1016/j.jcp.2006.02.014
25. Svärd M, Nordström J (2008) A stable high-order finite difference scheme for the compressible Navier-Stokes equations: no-slip wall boundary conditions. J Comput Phys 227(10):4805–4824
26. Svärd M, Nordström J (2014) Review of summation-by-parts schemes for initial-boundary-value problems. J Comput Phys 268:17–38
27. Sweby PR (1984) High resolution schemes using flux limiters for hyperbolic conservation laws. SIAM J Numer Anal 21(5):995–1011
28. Toro EF (1999) Riemann solvers and numerical methods for fluid dynamics: a practical introduction, 2nd edn. Springer, Berlin
29. Toro EF, Spruce M, Speares W (1994) Restoration of the contact surface in the HLL-Riemann solver. Shock Waves 4:25–34. http://dx.doi.org/10.1007/BF01414629
30. van Leer B (1979) Towards the ultimate conservative difference scheme. V – a second-order sequel to Godunov's method. J Comput Phys 32:101–136. doi:10.1016/0021-9991(79)90145-1

Part II
Scalar Transport Problems

Chapter 5
Linear Transport Under Uncertainty

The aim of this chapter, based on [18], is to present accurate and stable numerical schemes for the solution of a class of linear diffusive transport problems. The advection-diffusion equation subject to uncertain viscosity with known statistical description is represented by a spectral expansion in the stochastic dimension. The gPC framework and the stochastic Galerkin method are used to obtain an extended system which is analyzed to find discretization constraints on monotonicity, stiffness and stability. A comparison of stochastic Galerkin versus methods based on repeated evaluations of deterministic solutions, such as stochastic collocation, is provided but this is not our primary focus. However, we do include a few examples on relative performance and numerical properties with respect to monotonicity requirements and convergence to steady-state, to encourage the use of stochastic Galerkin methods.

Special care is exercised to ensure that the stochastic Galerkin projection results in a system with a positive semidefinite diffusion matrix. The sign of the eigenvalues of a pure advection problem is not a stumbling block as long as the boundary conditions are properly adjusted to match the number of ingoing characteristics, as shown in [7]. Unlike the case of stochastic advection, the sign of the eigenvalues of the diffusion matrix of the advection-diffusion problem is crucial. A negative eigenvalue leads to the growth of the solution norm and hence numerical instability. The source of the growth is in the volume term, and no treatment of the boundary conditions can eliminate it.

Advection-diffusion problems with uncertainty have been investigated by several authors. Ghanem and Dham [5] considered a lognormal diffusion coefficient in a multiphase porous medium problem. Le Maître et al. investigated a set of Navier-Stokes problems, resulting in coupled sets of advection-diffusion equations with uncertain diffusion [13]. Wan et al. investigated the advection-diffusion equation in two dimensions with random transport velocity [27], and the effect of long-term time integration of flow problems with gPC methods [26]. Xiu and Karniadakis

M.P. Pettersson et al., *Polynomial Chaos Methods for Hyperbolic Partial Differential Equations*, Mathematical Engineering,
DOI 10.1007/978-3-319-10714-1_5

studied the Navier-Stokes equations with various stochastic boundary conditions [29], as well as steady-state problems with random diffusivity [28]. We extend the work by previous authors through analysis of the numerical method used for the stochastic Galerkin problem, e.g., investigating monotonicity and stability requirements and convergence to steady-state.

The stochastic advection-diffusion equation and the stochastic Galerkin formulation are presented in Sect. 5.1. We consider an uncertain diffusion coefficient μ which is replaced by a stochastic Galerkin matrix, whose eigenvalues will determine the rate of diffusion of the solution. Different basis functions and estimates of the eigenvalues of the diffusion matrix are given in Sect. 5.2. These eigenvalues and their relation to the deterministic velocity determine the dynamics of the solution. They also add restrictions on the numerical solution methods.

We prove well-posedness of the stochastic Galerkin problem in Sect. 5.3. This proof serves to demonstrate that we have chosen proper boundary conditions. The impact of the eigenvalues of the stochastic Galerkin diffusion matrix on the numerical method is demonstrated in Sect. 5.4, where monotonicity requirements for the numerical solution are discussed. In Sect. 5.5, we investigate the time-step limitations of the numerical schemes using the von Neumann analysis for a periodic case. The von Neumann analysis is not applicable for non-periodic solutions, hence we use summation-by-parts operators to show stability for the non-periodic case. We consider a spatially constant as well as a spatially varying diffusion to demonstrate different features of the SBP framework. Section 5.5 also includes analysis regarding the convergence rate of the steady-state problem. Numerical results are then presented in Sect. 5.6.

5.1 Problem Definition

Let $(\Omega, \mathscr{F}, \mathscr{P})$ be a suitable probability space with the set of elementary events Ω and probability measure $\mathscr{P}$ defined on the σ-algebra $\mathscr{F}$. Let $\xi(\omega)$, $\omega \in \Omega$, be a random variable defined on this space. Consider the following mixed hyperbolic-parabolic stochastic PDE defined on $(0,1) \times [0,T]$ which holds $\mathscr{P}$-almost surely in Ω,

$$\begin{aligned} \frac{\partial u}{\partial t} + v\frac{\partial u}{\partial x} &= \frac{\partial}{\partial x}\left(\mu(x,\xi)\frac{\partial u}{\partial x}\right), \\ u(0,t,\xi) &= g_0(t,\xi), \qquad (5.1) \\ \frac{\partial u(x,t,\xi)}{\partial x}\Big|_{x=1} &= g_1(t,\xi), \\ u(x,0,\xi) &= u_{init}(x,\xi). \qquad (5.2) \end{aligned}$$

Here the velocity $v > 0$ is a deterministic scalar and the diffusion $\mu(x,\xi) > \mu_0 > 0$ is a finite variance random field. As a special case of (5.1), we consider the case of

$\mu(\xi)$ being constant in space, i.e., homogeneous and also dependent on the uncertain parameter ξ, but with the same initial and boundary conditions.

In what follows, we approximate the stochastic solution $u(x, t, \xi)$ using a gPC expansion in the random space. We use the stochastic Galerkin method and compare with the stochastic collocation method. Our objective is then to explore the stability, stiffness and monotonicity requirements associated with the numerical solution of the resulting coupled system of equations.

5.1.1 Uncertainty and Solution Procedure

We will consider the case where μ has a uniform probability distribution and thus bounded range, and the case where μ takes a lognormal distribution, a common model in geophysics applications such as transport in porous media [3]. For other distributions, we assume that the diffusion coefficient $\mu(\xi)$ has the cumulative distribution function F. One may parameterize the uncertainty with a uniform random variable ξ, defined on the interval $[-1, 1]$ with constant probability density 0.5, denoted $\xi \sim \mathscr{U}[-1, 1]$. Then we get the expression

$$\mu(\xi) = F^{-1}\left(\frac{\xi + 1}{2}\right), \tag{5.3}$$

which holds for general distributions F when F^{-1} is defined. For the cases of interest here, F^{-1} is a linear function in the case of a uniform μ. In the case of a lognormal μ, we will alternatively represent μ in terms of the Hermite polynomial chaos expansion in a Gaussian random variable.

In the context of a stochastic Galerkin solution of (5.1), we expand the solution $u(x, t, \xi)$ with respect to a gPC basis $\{\psi_k(\xi)\}_{k=0}^{\infty}$. Legendre and Hermite orthogonal polynomials are both used in the numerical experiments. In the computations, we need to use a basis with finite cardinality, as indicated before. Hence, we truncate the gPC basis $\{\psi_k(\xi)\}_{k=0}^{\infty}$ to exactly represent polynomials up to order M,

$$u^M(x, t, \xi) = \sum_{k=0}^{M} u_k(x, t)\psi_k(\xi), \tag{5.4}$$

where $\{\psi_k(\xi)\}_{k=0}^{M}$ is the set of gPC basis functions of maximum order M.

5.1.2 Stochastic Galerkin Projection

The unknown coefficients $u_k(x, t)$ are computed through a Galerkin projection onto the subspace spanned by the basis $\{\psi_k(\xi)\}_{k=0}^{M}$. Specifically, the truncated series (5.4) is inserted into (5.1) and multiplied by each one of the basis functions $\{\psi_k(\xi)\}_{k=0}^{M}$. The resulting expression is integrated with respect to the probability measure $\mathscr{P}$ over the stochastic domain. This leads to a coupled linear system of deterministic PDE's of the form

$$\frac{\partial u_k}{\partial t} + v\frac{\partial u_k}{\partial x} = \sum_{j=0}^{M} \frac{\partial}{\partial x}\left(\langle \mu\psi_j\psi_k\rangle \frac{\partial u_j}{\partial x}\right), \quad k = 0,\ldots,M,$$
$$u_k(0,t) = (g_0)_k, \quad k = 0,\ldots,M,$$
$$\frac{\partial u_k(x,t)}{\partial x}|_{x=1} = (g_1)_k, \quad k = 0,\ldots,M,$$
$$u_k(x,0) = (u_{init})_k, \quad k = 0,\ldots,M, \tag{5.5}$$

where the orthogonality of the basis functions $\{\psi_k(\xi)\}_{k=0}^{M}$ has been used to cancel terms. Here, $(g_0)_k$, $(g_1)_k$ and $(u_{init})_k$ are obtained by the projection of the left and right boundary data and the initial function on basis polynomial $\psi_k(\xi)$, $k = 0,\ldots,M$. In the sequel we use a compact notation to represent the system (5.5). Let $\boldsymbol{u}^M \equiv (u_0\ u_1\ \ldots\ u_M)^T$ be the vector of gPC coefficients in (5.4). Then, the system (5.5) can be equivalently written as

$$\frac{\partial \boldsymbol{u}^M}{\partial t} + \boldsymbol{V}\frac{\partial \boldsymbol{u}^M}{\partial x} = \frac{\partial}{\partial x}\left(\boldsymbol{B}(x)\frac{\partial \boldsymbol{u}^M}{\partial x}\right), \tag{5.6}$$
$$\boldsymbol{u}^M(0,t) = \boldsymbol{g}_0^M(t),$$
$$\frac{\partial \boldsymbol{u}^M(x,t)}{\partial x}|_{x=1} = \boldsymbol{g}_1^M(t),$$
$$\boldsymbol{u}^M(x,0) = \boldsymbol{u}_{init}^M(x), \tag{5.7}$$

where $\boldsymbol{V} = diag(v)$ and the matrix $\boldsymbol{B}$ is defined by

$$[\boldsymbol{B}(x)]_{jk} = \langle \mu(x,\xi)\psi_j\psi_k\rangle \quad j,k = 0,\ldots,M. \tag{5.8}$$

We will frequently refer to the case of spatially independent $\mu(\xi)$. Then, (5.6) can be simplified to

$$\frac{\partial \boldsymbol{u}^M}{\partial t} + \boldsymbol{V}\frac{\partial \boldsymbol{u}^M}{\partial x} = \boldsymbol{B}\frac{\partial^2 \boldsymbol{u}^M}{\partial x^2}. \tag{5.9}$$

With the gPC expansion of the diffusion coefficient, $\mu(x,\xi) = \sum_{k=0}^{\infty}\mu_k(x)\psi_k(\xi)$, (5.8) can be rewritten as

$$[\boldsymbol{B}]_{ij} = \langle \mu\psi_i\psi_j\rangle = \sum_{k=0}^{\infty}\mu_k(x)\langle \psi_i\psi_j\psi_k\rangle, \quad i,j = 0,\ldots,M. \tag{5.10}$$

For the basis functions that will be used in this chapter, all triple (inner) products $\langle\psi_i\psi_j\psi_k\rangle$ satisfy

$$\langle\psi_i\psi_j\psi_k\rangle = 0, \qquad \text{for } k > 2M \text{ and } i, j \leq M. \tag{5.11}$$

Explicit formulas for $\langle\psi_i\psi_j\psi_k\rangle$ for Hermite and Legendre polynomials can be found in [1, 25]. Hence, using (5.11), (5.10) may be simplified to

$$[\boldsymbol{B}]_{ij} = \sum_{k=0}^{2M} \mu_k(x)\langle\psi_i\psi_j\psi_k\rangle, \quad i, j = 0, \ldots, M. \tag{5.12}$$

The entries of $\boldsymbol{B}$ can thus be evaluated as finite sums of triple products that can be computed exactly. Moreover, since $[\boldsymbol{B}]_{ij} = \langle\mu\psi_i\psi_j\rangle = \langle\mu\psi_j\psi_i\rangle = [\boldsymbol{B}]_{ji}$, it follows that $\boldsymbol{B}$ is symmetric.

It is essential that the matrix $\boldsymbol{B}$ always be positive definite when it is derived from a well-defined $\mu(\xi) > 0$. This holds as a consequence of the following proposition. The proof of the proposition follows closely that of the positive (negative) definiteness of the advection matrix of Theorem 2.1 in [7] and Theorem 3.1 in [30]. However, here we also emphasize the importance of a suitable polynomial chaos approximation of $\boldsymbol{B}$, since in this case negative eigenvalues would lead to instability of the numerical method.

Proposition 5.1. *The diffusion matrix $\boldsymbol{B}$ given by (5.12) derived from any $\mu(\xi)$ satisfying $\mu(\xi) \geq 0\mathscr{P}$-almost surely in Ω, has non-negative eigenvalues.*

Proof. For any order M of gPC expansion and any vector $\boldsymbol{u}^M \in \mathbb{R}^{M+1}$,

$$(\boldsymbol{u}^M)^T \boldsymbol{B}\boldsymbol{u}^M = \sum_{i=0}^{M}\sum_{j=0}^{M} u_i u_j \sum_{k=0}^{2M}\langle\psi_i\psi_j\psi_k\rangle\mu_k = \sum_{i=0}^{M}\sum_{j=0}^{M} u_i u_j \langle\psi_i\psi_j\mu\rangle =$$

$$= \int_{\Omega}\left(\sum_{i=0}^{M} u_i\psi_i\right)^2 \mu(\xi)d\mathscr{P}(\xi) \geq 0. \tag{5.13}$$

Remark 5.1. The above proposition does not hold for the order M approximation $\tilde{\mu}(\xi) = \sum_{k=0}^{M} \mu_k\psi_k(\xi)$. The second equality of (5.13) relies on substituting the gPC expansion of μ of order $2M$ with the full gPC expansion of μ. This substitution is valid following (5.11), but it would not be valid for the order M gPC approximation of μ. In the latter case, the resulting $\boldsymbol{B}$ may have negative eigenvalues, thus ruining the stability of the discrete approximation of (5.6). Therefore, the $2M$ order of gPC expansion of μ is crucial. Figure 5.1 illustrates this for the case of a lognormal $\mu(\xi) = \exp(\xi)$ with $\xi \sim \mathcal{N}(0, 1)$.

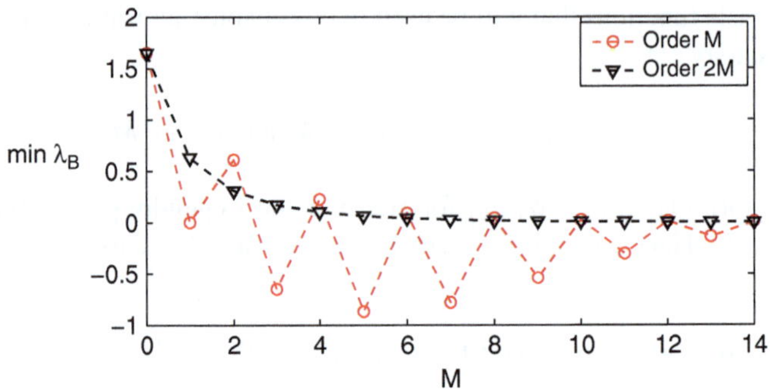

Fig. 5.1 Minimum $\lambda_{\boldsymbol{B}}$ for $\mu = \exp(\xi)$. Here $[\boldsymbol{B}^{(M)}]_{ij} = \sum_{k=0}^{M}\langle\psi_i\psi_j\psi_k\rangle\mu_k$ and $[\boldsymbol{B}^{(2M)}]_{ij} = \sum_{k=0}^{2M}\langle\psi_i\psi_j\psi_k\rangle\mu_k$, respectively, and $\{\psi_k(\xi)\}$ are the Hermite polynomials

5.1.3 Diagonalization of the Stochastic Galerkin System

In order to reduce the computational cost, it is advantageous to diagonalize the stochastic Galerkin systems whenever possible. If this is indeed possible, exact or numerical diagonalization can be done as a preprocessing step, followed by the numerical solution of $M+1$ scalar advection-diffusion problems with different, but strictly positive, viscosity $\mu((\lambda_{\boldsymbol{B}})_j)$, where $(\lambda_{\boldsymbol{B}})_j$ are the eigenvalues of $\boldsymbol{B}$, $j = 0,\ldots,M$. The system (5.6) can be diagonalized under certain conditions, which we elaborate on next. Assuming, for a moment, that $\boldsymbol{B}(x) = \boldsymbol{W}\boldsymbol{\Lambda}_{\boldsymbol{B}}(x)\boldsymbol{W}^T$, i.e. that the eigenvectors $\boldsymbol{W}$ of $\boldsymbol{B}(x)$ are not spatially dependent, then the system (5.6) can be diagonalized. Multiplying (5.6) from the left by $\boldsymbol{W}^T$ and letting $\tilde{\boldsymbol{u}}^M = \boldsymbol{W}^T\boldsymbol{u}^M$, we get the diagonalized system

$$\frac{\partial\tilde{\boldsymbol{u}}^M}{\partial t} + \boldsymbol{V}\frac{\partial\tilde{\boldsymbol{u}}^M}{\partial x} = \frac{\partial}{\partial x}\left(\boldsymbol{\Lambda}_{\boldsymbol{B}}(x)\frac{\partial\tilde{\boldsymbol{u}}^M}{\partial x}\right).$$

When the stochastic and space-dependent components of $\mu(x,\xi)$ can be factorized or only occur in separate terms of a sum, $\boldsymbol{B}(x)$ can be diagonalized. That is, for general nonlinear functions f, g and h, and $\mu(x,\xi) = f(x)g(\xi) + h(x)$, we have

$$\boldsymbol{B}(x) = f(x)\boldsymbol{W}\boldsymbol{\Lambda}_g\boldsymbol{W}^T + h(x) = \boldsymbol{W}\left(f(x)\boldsymbol{\Lambda}_g + h(x)\boldsymbol{I}\right)\boldsymbol{W}^T = \boldsymbol{W}\boldsymbol{\Lambda}_{\boldsymbol{B}}(x)\boldsymbol{W}^T,$$

where $\boldsymbol{\Lambda}_{\boldsymbol{B}}(x) = f(x)\boldsymbol{\Lambda}_g + h(x)\boldsymbol{I}$, $\boldsymbol{\Lambda}_g$ is a diagonal matrix, $\boldsymbol{W}$ is the eigenvector matrix of the eigenvalue decomposition of $[\boldsymbol{B}_g]_{ij} = \langle g\psi_i\psi_j\rangle$, and $\boldsymbol{I}$ is the identity matrix. The only requirement on f, g, and h is that the resulting $\mu(x,\xi)$ be positive for all ξ, and bounded in the $L^2(\Omega,\mathscr{P})$ norm.

Notice that the form $\mu(x,\xi) = f(x)g(\xi) + h(x)$ has a given distribution throughout the domain, but not necessarily with the parameters of the distribution being constant. For instance, with $\mu = c_1(x) + c_2(x)\exp(\xi)$ and $\xi \sim \mathscr{N}(0,1)$, the viscosity is lognormal for all x but with spatially varying statistics and diagonalization. However, for the general case $\mu(x, \xi_1, \ldots, \xi_d) = \exp(G(x, \xi_1, \ldots, \xi_d))$, with G being a multivariate Gaussian field, diagonalization is not possible.

For the general case of any empirical distribution with simultaneous spatial and stochastic variation diagonalization is not possible. Then we solve the full stochastic Galerkin system, analysis of which is described in the following sections. We also present results on the diagonalizable case, since this allows a very direct comparison to the stochastic collocation techniques, presented next.

5.2 The Eigenvalues of the Diffusion Matrix *B*

In the analysis of the mathematical properties and the numerical scheme, e.g., well-posedness, monotonicity, stiffness and stability, we need estimates of the eigenvalues of $\boldsymbol{B}$. We may express

$$\boldsymbol{B} = \sum_{k=0}^{\infty} \mu_k \boldsymbol{C}_k, \tag{5.14}$$

where μ_k's are the polynomial chaos coefficients of $\mu(\xi)$ and $[\boldsymbol{C}_k]_{ij} = \langle \psi_i \psi_j \psi_k \rangle$.

5.2.1 General Bounds on the Eigenvalues of *B*

Some eigenvalue estimates pertain to all gPC expansions, independent of the actual choice of stochastic basis functions. For example, in cases where $\mu(\xi)$ is bounded within an interval of the real line, the eigenvalues of the viscosity matrix $\boldsymbol{B}$ can essentially be bounded from above and below by the upper and lower interval boundaries of possible values of μ, respectively. More generally, for any countable basis $\{\psi_k(\xi)\}_{k=0}^{\infty}$ of $L_2(\Omega, \mathscr{P})$, by Theorem 2 of [22], it follows that there is a bound on the set $\{(\lambda_{\boldsymbol{B}})_j\}_{j=0}^{M}$ of the eigenvalues of $\boldsymbol{B}$, given by

$$(\lambda_{\boldsymbol{B}})_j \in conv(spect(\mu(\xi))) = [\mu_{min}, \mu_{max}], \tag{5.15}$$

where $conv$ denotes the convex hull, and the spectrum $spect$ of $\mu(\xi)$ is the essential range, i.e., the set of all possible values (measurable) μ can attain. For a more general exposition and for cases where μ is not confined to a convex region, we refer the interested reader to [22]. Here, we only consider μ in intervals of finite or infinite length (convex sets), and do not consider degenerate sets or single point values. Following (5.15), for bounded μ such as uniformly distributed viscosity, the eigenvalues $(\lambda_{\boldsymbol{B}})_j$ will be restricted to an interval for all orders M of gPC expansion. We expect that the order of polynomial chaos expansion has a limited impact on system properties such as monotonicity and stiffness for these cases, as demonstrated in Sect. 5.5.3. For unbounded μ (e.g., lognormal distribution) there is no upper bound on the eigenvalues of $\boldsymbol{B}$ and the system properties change with the order of gPC, also shown in Sect. 5.5.3.

5.2.2 Legendre Polynomial Representation

When the viscosity μ is given by $\mu = \mu_0 + \hat{\sigma}\xi$, $\xi \sim \mathscr{U}[-1,1]$ and $\hat{\sigma}$ is a deterministic scaling factor, only the first two Legendre polynomials are needed to represent μ exactly, that is $\mu = \mu_0\psi_0 + \hat{\sigma}/\sqrt{3}\psi_1$. Then, the stochastic Galerkin projection yields a matrix $\boldsymbol{B}$ of the form

$$[\boldsymbol{B}]_{jk} = \langle \mu\psi_j\psi_k \rangle = \mu_0 I + \mu_1 \boldsymbol{C}_1, \quad j,k = 0,\ldots,M,$$

where the eigenvalues of $\boldsymbol{C}_1$ ($[C_1]_{i,j} = \langle \psi_1\psi_i\psi_j \rangle$) are given by the Gauss-Legendre quadrature nodes scaled by $\sqrt{3}$. The scaling factor is due to the normalization performed to obtain unit-valued inner double products of the Legendre polynomials. This result follows from the fact that the eigenvalues of the matrix with (i,j) entries defined by $\langle \xi\psi_i\psi_j \rangle$ are the same as those of the Jacobi matrix corresponding to the three-term recurrence of the Legendre polynomials. Thus, they are equal to the Gauss-Legendre quadrature nodes (see e.g. [6, 25] for further details on this assertion).

The Gauss-Legendre nodes are located in the interval $[-1,1]$, from which it follows that $(\lambda_{\boldsymbol{B}})_j \in [\mu_0 - \hat{\sigma}, \mu_0 + \hat{\sigma}]$. Note that this holds exactly only for a uniformly distributed μ; for non-uniform μ, the polynomial expansion would result in a matrix series representation of $\boldsymbol{B}$ of the form (5.14), where the matrices $\boldsymbol{C}_k$ are nonzero also for $k > 1$.

5.2.3 Hermite Polynomial Representation

Representing the uncertainty of the input parameters with an orthogonal polynomial basis whose weight function does not match the probability measure of the input parameters may lead to poor convergence rates [29]. However, problems where the inputs are functions of Gaussian variables may be represented by gPC expansions in the Hermite polynomials with a weight function matching the Gaussian measure. For instance, lognormal random processes can effectively be represented by Hermite polynomial chaos expansion (see e.g., [4]). Let

$$\mu(\xi) = c_1 + c_2 e^{\xi}, \quad c_1, c_2 \geq 0, \xi \sim \mathscr{N}(0,1). \tag{5.16}$$

Then, the Hermite polynomial chaos coefficients of μ are given by

$$\mu_j = \frac{c_2 e^{1/2}}{\sqrt{j!}}, \qquad j \geq 1. \tag{5.17}$$

The inner triple products of Hermite polynomials are given by

$$\langle \psi_i \psi_j \psi_k \rangle = \begin{cases} \frac{\sqrt{i!j!k!}}{(s-i)!(s-j)!(s-k)!} & s \text{ integer}, i, j, k \leq s \\ 0 & \text{otherwise}, \end{cases} \tag{5.18}$$

with $s = (i + j + k)/2$.

Applying Proposition 5.1 of Sect. 5.1.2 to the lognormal μ in (5.16), it follows that the eigenvalues of $\boldsymbol{B}$ are bounded below by c_1. The largest eigenvalue grows with the order M of gPC expansion. Since the entries of $\boldsymbol{B}$ are non-negative due to (5.17) and (5.18), by the Gershgorin's circle theorem, the largest eigenvalue is bounded by the maximum row (column) sum of $\boldsymbol{B}$. This gives an estimate of the stiffness of the problem, where a problem is considered stiff when the time-step required for stability is much smaller than that required for accuracy [23].

5.3 Boundary Conditions for Well-Posedness

A problem is *well-posed* if a solution exists, is unique and depends continuously on the problem data. Boundary conditions that lead to a bounded energy are necessary for well-posedness. For hyperbolic stochastic Galerkin systems, boundary conditions have been derived in [7] for the linear wave equation and in [19] for the nonlinear case of Burgers' equation. Given the setting of (5.6), we derive the energy equation by multiplying $(\boldsymbol{u}^M)^T$ with the first equation in (5.6) and integrating over the spatial extent of the problem. More specifically,

$$\int_0^1 (\boldsymbol{u}^M)^T \frac{\partial \boldsymbol{u}^M}{\partial t} dx + \int_0^1 (\boldsymbol{u}^M)^T \boldsymbol{V} \frac{\partial \boldsymbol{u}^M}{\partial x} dx = \int_0^1 (\boldsymbol{u}^M)^T \frac{\partial}{\partial x} \left(\boldsymbol{B}(x) \frac{\partial \boldsymbol{u}^M}{\partial x} \right) dx, \tag{5.19}$$

which can be compactly written as

$$\begin{aligned} &\frac{\partial \|\boldsymbol{u}^M\|^2}{\partial t} + 2 \int_0^1 \frac{\partial (\boldsymbol{u}^M)^T}{\partial x} \boldsymbol{B}(x) \frac{\partial \boldsymbol{u}^M}{\partial x} dx \\ &= \left[(\boldsymbol{u}^M)^T \boldsymbol{V} \boldsymbol{u}^M - 2 (\boldsymbol{u}^M)^T \boldsymbol{B}(x) \frac{\partial \boldsymbol{u}^M}{\partial x} \right]_{x=0} \\ &\quad - \left[(\boldsymbol{u}^M)^T \boldsymbol{V} \boldsymbol{u}^M - 2 (\boldsymbol{u}^M)^T \boldsymbol{B}(x) \frac{\partial \boldsymbol{u}^M}{\partial x} \right]_{x=1}. \end{aligned} \tag{5.20}$$

Proposition 5.2. *The problem (5.6) is well-posed in the sense of Definition 1.1.*

Proof. We consider homogeneous boundary conditions, i.e., let $\boldsymbol{g}_0^M = \boldsymbol{g}_1^M = \boldsymbol{0}$ in (5.7). Notice that the right-hand side of (5.20) is negative for the choice of boundary conditions in (5.7), hence leading to a bounded energy norm of solution u in time. Uniqueness follows directly from the energy estimate by replacing the solution by the difference between two solutions $\boldsymbol{u}^M$ and $\boldsymbol{v}^M$ and noticing that the norm of the difference is non-increasing with time, thus $\boldsymbol{u}^M \equiv \boldsymbol{v}^M$. The problem is parabolic with full-rank $\boldsymbol{B}$ and the correct number of boundary conditions. This implies the existence of the solution. Therefore, the problem (5.6) (and also (5.1)) is well-posed.

5.4 Monotonicity of the Solution

In this section we use a *normal modal analysis* technique [8] to derive the necessary conditions for the monotonicity of the steady-state solution of the system of (5.9) with spatially constant, but random, viscosity. We provide these conditions for second- and fourth-order discretization operators.

5.4.1 Second-Order Operators

With standard second-order central differences and a uniform grid, the semidiscrete representation of (5.9) for the steady-state limit reads

$$\boldsymbol{V}\frac{\boldsymbol{u}_{i+1}^M - \boldsymbol{u}_{i-1}^M}{2\Delta x} = \boldsymbol{B}\frac{\boldsymbol{u}_{i+1}^M - 2\boldsymbol{u}_i^M + \boldsymbol{u}_{i-1}^M}{\Delta x^2}, \tag{5.21}$$

where $\boldsymbol{u}_i^M$ denotes the sub-vector of the vector of the discretized solution $\vec{\boldsymbol{u}}$ at the grid point i in space. This is a system of difference equations with a solution of the form

$$\boldsymbol{u}_i^M = \boldsymbol{y}^M \kappa^i, \tag{5.22}$$

for some scalar κ and vector $\boldsymbol{y}^M \in \mathbb{R}^{M+1}$ to be determined. By inserting (5.22) into (5.21) we arrive at the eigen-problem

$$\left[\frac{\Delta x(\kappa^2 - 1)}{2}\boldsymbol{V} - (\kappa - 1)^2\boldsymbol{B}\right]\boldsymbol{y}^M = \boldsymbol{0}, \tag{5.23}$$

whose non-trivial solution is obtained by requiring

$$\det\left(\frac{\Delta x(\kappa^2-1)}{2}\boldsymbol{V}-(\kappa-1)^2\boldsymbol{B}\right)=0. \tag{5.24}$$

The spectral decomposition of the symmetric positive definite matrix $\boldsymbol{B}$, i.e., $\boldsymbol{B}=\boldsymbol{W}\boldsymbol{\Lambda}_{\boldsymbol{B}}\boldsymbol{W}^T$, inserted into (5.24) leads to

$$v\Delta x(\kappa_j^2-1)-(\lambda_{\boldsymbol{B}})_j(\kappa_j-1)^2=0, \qquad j=0,\ldots,M. \tag{5.25}$$

The solution to (5.25) is

$$\kappa_j=1 \text{ or } \frac{2+\theta_j}{2-\theta_j}, \qquad j=0,\ldots,M, \tag{5.26}$$

where $\theta_j=\frac{v\Delta x}{(\lambda_{\boldsymbol{B}})_j}$.

For a monotonic solution $\vec{\boldsymbol{u}}$, we must have $\kappa_j\geq 0$, which demands a mesh such that

$$\mathrm{Re}_{\mathrm{mesh}}=\max_j \theta_j\leq 2. \tag{5.27}$$

In the case of stochastic collocation, each realization will have a different mesh Reynolds number $\mathrm{Re}_{\mathrm{mesh}}$ based on the value of $\mu(\xi)$. In combination with the Courant-Friedrichs-Lewy (CFL) restriction on the time-step Δt, this allows for larger time-steps for simulations corresponding to large values of $\mu(\xi)$, but forces small ones for small $\mu(\xi)$.

The importance of the mesh Reynolds number is illustrated in Fig. 5.2. A step function initially located at $x=0.2$ is transported to the right and is increasingly smeared by viscosity $\mu\sim\mathscr{U}[0.05,0.15]$. The mean value is monotonically decreasing, but this property is clearly not preserved by numerical schemes that do not satisfy the mesh Reynolds number requirement. It also has the effect of erroneously predicting the location of the variance peaks.

When $\boldsymbol{B}$ can be diagonalized, the solution statistics are functions of linear combinations of scalar advection-diffusion solutions with viscosity given by the eigenvalues $(\lambda_{\boldsymbol{B}})_j$. Then, there is a *local* mesh Reynolds number $(\mathrm{Re}_{\mathrm{mesh}})_j=\theta_j$ for each eigenvalue $(\lambda_{\boldsymbol{B}})_j$, and a *global* mesh Reynolds number $\mathrm{Re}_{\mathrm{mesh}}$ defined by (5.27). $\mathrm{Re}_{\mathrm{mesh}}$ is defined also for cases when $\boldsymbol{B}$ cannot be diagonalized. If the global mesh Reynolds number for the Galerkin system $\mathrm{Re}_{\mathrm{mesh}}>2$, but the local mesh Reynolds number $(\mathrm{Re}_{\mathrm{mesh}})_j<2$ for some instances of the scalar advection-diffusion equation after diagonalization, the lack of monotonicity may not be obvious in the statistics, since these are affected by averaging effects from all scalar solutions. Hence, the lack of monotonicity of the mean solution is more obvious if $(\mathrm{Re}_{\mathrm{mesh}})_j>2$ for all $j=0,\ldots,M$. This is shown in Fig. 5.3 with $\mu\sim\mathscr{U}[0.14,0.16]$ for $\mathrm{Re}_{\mathrm{mesh}}=3$ (and $(\mathrm{Re}_{\mathrm{mesh}})_j>2$, $j=0,\ldots,M$) and $\mathrm{Re}_{\mathrm{mesh}}=1$, respectively.

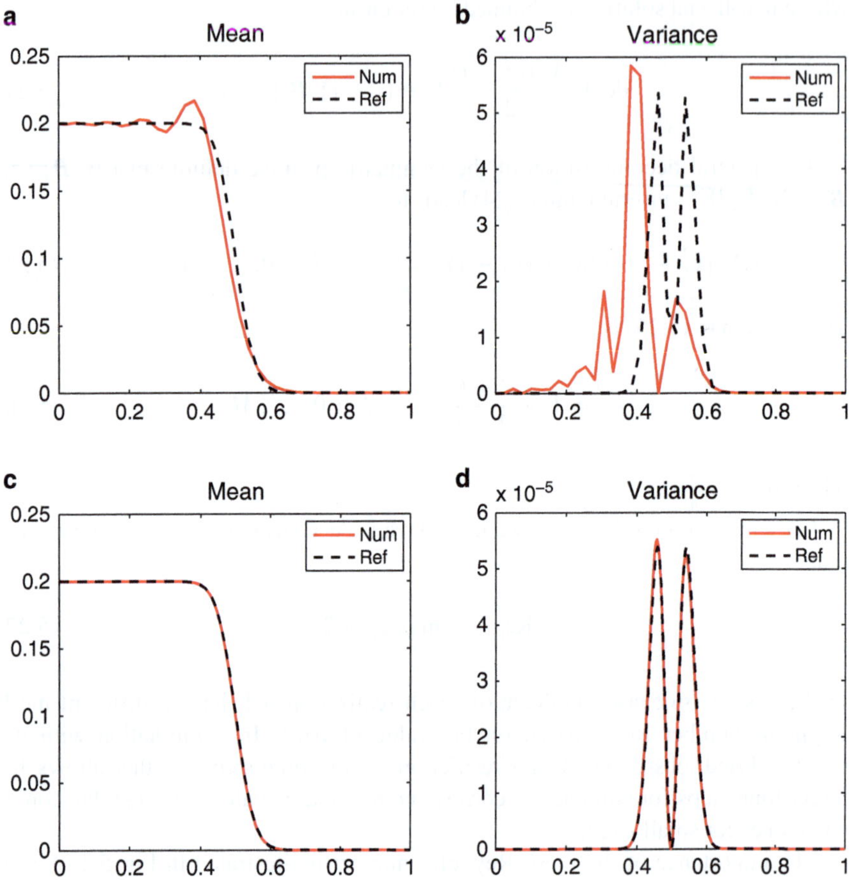

Fig. 5.2 Solution statistics at $t = 0.01$ using stochastic Galerkin with $M = 4$ for diffusion of a moving step function, $u(x,t,\xi) = \rho_0 \text{erfc}\left((x-(x_0+v(t+\tau)))/\sqrt{(4\mu(\xi)(t+\tau))}\right)$, $\mu(\xi) \sim \mathscr{U}[0.05, 0.15]$, $\rho_0 = 0.1$, $\tau = 0.005$, $x_0 = 0.2$, and $v = 1$. Here, m denotes the number of spatial grid points. (**a**) $m = 40$, $\text{Re}_{\text{mesh}} = 14$. (**b**) $m = 40$, $\text{Re}_{\text{mesh}} = 14$. (**c**) $m = 300$, $\text{Re}_{\text{mesh}} = 1.9$. (**d**) $m = 300$, $\text{Re}_{\text{mesh}} = 1.9$

Remark 5.2. The condition on the mesh Reynolds number is no longer present with an upwind scheme, expressed as a central scheme with a certain amount of artificial dissipation. To see this, let the diagonalized scheme with artificial dissipation be given by

$$V\frac{\boldsymbol{u}_{i+1}-\boldsymbol{u}_{i-1}}{2\Delta x} - \boldsymbol{\Lambda_B}\frac{\boldsymbol{u}_{i+1}-2\boldsymbol{u}_i+\boldsymbol{u}_{i-1}}{\Delta x^2} = \alpha(\boldsymbol{u}_{i+1}-2\boldsymbol{u}_i+\boldsymbol{u}_{i-1}).$$

The choice $\alpha = v/(2\Delta x)$ leads to upwinding. With the ansatz (5.22), we get $\kappa_j = 1$ or $\kappa_j = 1 + v\Delta x/(\lambda_{\boldsymbol{B}})_j$ for $j = 0,\ldots,M$. This shows that the solution is oscillation free independent of the mesh Reynolds number. However, the upwinding adversely affects the accuracy of the solution.

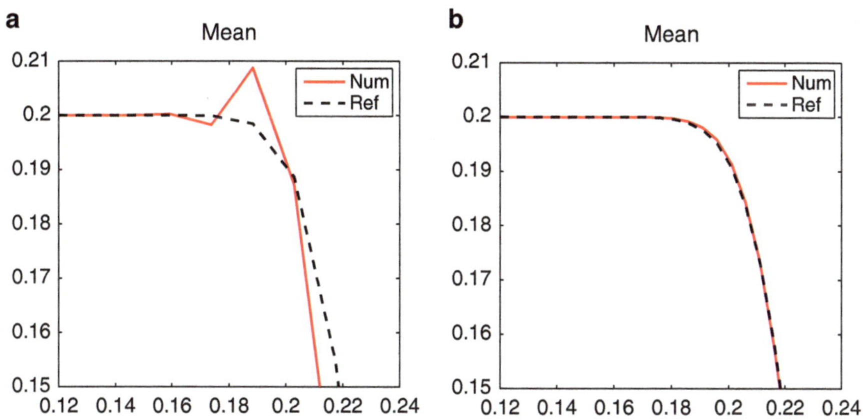

Fig. 5.3 Mean solution at $t = 0.001$ for diffusion of a moving step function, $M = 4$. (**a**) $\mathrm{Re}_{\mathrm{mesh}} = 3$, $m = 70$. (**b**) $\mathrm{Re}_{\mathrm{mesh}} = 1$, $m = 200$

5.4.2 Fourth-Order Operators

With fourth-order central differences, the semidiscrete representation of (5.9) for the steady-state limit is given by

$$\begin{aligned} &\boldsymbol{V}\frac{-\boldsymbol{u}_{i+2}^{M} + 8\boldsymbol{u}_{i+1}^{M} - 8\boldsymbol{u}_{i-1}^{M} + \boldsymbol{u}_{i-2}^{M}}{12\Delta x} \\ &= \boldsymbol{B}\frac{-\boldsymbol{u}_{i+2}^{M} + 16\boldsymbol{u}_{i+1}^{M} - 30\boldsymbol{u}_{i}^{M} + 16\boldsymbol{u}_{i-1}^{M} - \boldsymbol{u}_{i-2}^{M}}{12\Delta x^2}. \end{aligned} \tag{5.28}$$

Following the procedure of monotonicity analysis used for the second-order operators with the ansatz $\boldsymbol{u}_i^M = \boldsymbol{y}^M \kappa^i$ inserted in (5.28), we arrive at the eigen-problem

$$\left[(-\kappa^4 + 8\kappa^3 - 8\kappa + 1)\Delta x \boldsymbol{V} - (-\kappa^4 + 16\kappa^3 - 30\kappa^2 + 16\kappa - 1)\boldsymbol{B}\right]\boldsymbol{y}^M = \boldsymbol{0}. \tag{5.29}$$

One may verify that $\kappa = 1$ is a root of (5.29), just as in the case of second-order central differences. Using the spectral decomposition of $\boldsymbol{B}$ and factoring out $(\kappa - 1)$, we obtain the third-order equation

$$\left(1-\theta_j\right)\kappa_j^3 - \left(15 - 7\theta_j\right)\kappa_j^2 + \left(15 + 7\theta_j\right)\kappa_j - \left(1 + \theta_j\right) = 0, \tag{5.30}$$

for $j = 0, \ldots, M$. By Descartes' rule of signs, (5.30) has only positive roots $\kappa_j > 0$ for $0 < \theta_j < 1$. For $\theta_j > 1$, (5.30) has as at least one negative root. For $\theta_j = 1$, (5.30) reduces to a second-order equation with two positive roots. Hence, the monotonicity condition $\kappa_j \geq 0$ for the fourth-order operators is equivalent to the mesh Reynolds number bound

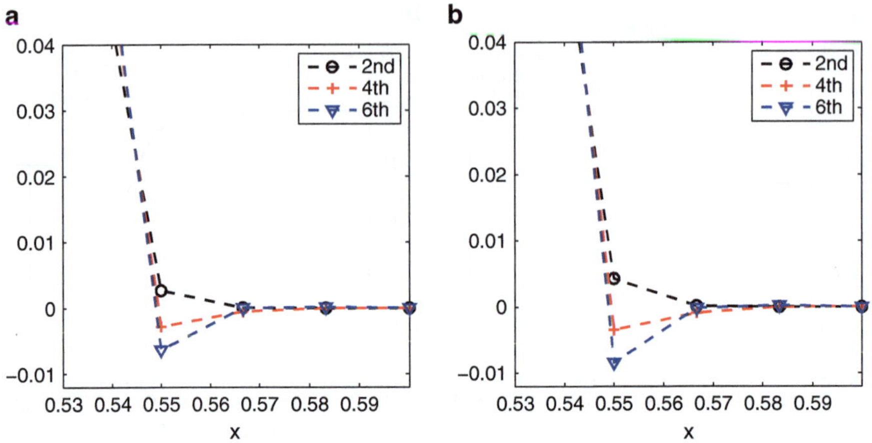

Fig. 5.4 Mean solution for diffusion of a moving step function after 40 time-steps, $M = 4$, $\mu \sim \mathscr{U}[0.0095, 0.0195]$, $m = 61$ spatial points and two different Re_{mesh}. The undershoot grows with the order of the operators. (**a**) $\text{Re}_{\text{mesh}} = 0.90$. (**b**) $\text{Re}_{\text{mesh}} = 1.90$

$$\text{Re}_{\text{mesh}} = \max_j \theta_j \leq 1. \tag{5.31}$$

Remark 5.3. The monotonicity analysis for sixth-order operators can be performed by following the method used for the fourth-order ones. The mesh Reynolds number monotonicity condition for sixth-order operators is $\text{Re}_{\text{mesh}} \leq \frac{2}{3}$.

Figure 5.4 depicts an initial step function after 40 time-steps, solved with second-, fourth- and sixth-order operators, respectively. The undershoots of the solutions tend to increase with the order of the scheme, which is inline with the restriction on Re_{mesh} that becomes more severe for higher-order operators.

5.5 Stability of the Semidiscretized Problem

A numerical scheme is *stable* if the semidiscrete problem with homogeneous boundary conditions leads to a bounded energy norm. A stable and consistent scheme converges by the Lax equivalence theorem. Our primary interest is the general case of non-periodic boundary conditions, but the well-known periodic case with spatially constant viscosity $\mu(\xi)$ is also included for comparison.

5.5.1 *The Initial Value Problem: von Neumann Analysis*

We consider the cases of second- and fourth-order accurate periodic versions of the central finite difference operators in [15], and show that the amplification factors have negative real parts, describing ellipses in the negative half-plane of the complex plane. The generalization to higher-order operators is straightforward.

5.5.1.1 Second-Order Operators

Assuming spatially constant μ and diagonalizing (5.9), and using the standard central difference discretization, we get

$$\frac{\partial \tilde{\boldsymbol{u}}_j^M}{\partial t} + \nu \frac{\tilde{\boldsymbol{u}}_{j+1}^M - \tilde{\boldsymbol{u}}_{j-1}^M}{2\Delta x} = \lambda_k \frac{\tilde{\boldsymbol{u}}_{j+1}^M - 2\tilde{\boldsymbol{u}}_j^M + \tilde{\boldsymbol{u}}_{j-1}^M}{(\Delta x)^2}. \tag{5.32}$$

We assume periodic boundary conditions and use the Fourier ansatz $\tilde{\boldsymbol{u}}_j^M = \hat{\boldsymbol{u}}^M e^{i\alpha\Delta x j}$, where α is the Fourier parameter. Then, with $\theta_k = \nu\Delta x/(2(\lambda_{\boldsymbol{B}})_k)$, (5.32) becomes

$$\begin{aligned}\frac{\partial \hat{\boldsymbol{u}}^M}{\partial t} &= -i\frac{\nu}{\Delta x}\frac{e^{i\alpha\Delta x} - e^{-i\alpha\Delta x}}{2i}\hat{\boldsymbol{u}}^M + \lambda_k \frac{e^{i\alpha\Delta x} - 2 + e^{-i\alpha\Delta x}}{(\Delta x)^2}\hat{\boldsymbol{u}}^M \\ &= -\frac{\nu}{\Delta x}\left[\sin(\alpha\Delta x)i + \frac{2}{\theta_k}(1 - \cos(\alpha\Delta x))\right]\hat{\boldsymbol{u}}^M.\end{aligned} \tag{5.33}$$

The coefficient of $\hat{\boldsymbol{u}}^M$ in the right-hand side of (5.33) is an expression of the form $f(\omega) = c_1\cos(\omega) + ic_2\sin(\omega) + c_3$, i.e., the parametrization of an ellipse in the complex plane. The real part is always non-positive due to the additive constant, so the spectrum is an ellipse in the negative half-plane.

5.5.1.2 Fourth-Order Operators

The fourth-order semidiscretization is given by

$$\begin{aligned}&\frac{\partial \tilde{\boldsymbol{u}}_j^M}{\partial t} + \nu \frac{-\tilde{\boldsymbol{u}}_{j+2}^M + 8\tilde{\boldsymbol{u}}_{j+1}^M - 8\tilde{\boldsymbol{u}}_{j-1}^M + \tilde{\boldsymbol{u}}_{j-1}^M}{12\Delta x} \\ &= \lambda_k \frac{-\tilde{\boldsymbol{u}}_{j+2}^M + 16\tilde{\boldsymbol{u}}_{j+1}^M - 30\tilde{\boldsymbol{u}}_j^M + 16\tilde{\boldsymbol{u}}_{j-1}^M - \tilde{\boldsymbol{u}}_{j-2}^M}{12(\Delta x)^2}.\end{aligned} \tag{5.34}$$

Using the Fourier ansatz, we have

$$\begin{aligned}\frac{\partial \hat{\boldsymbol{u}}^M}{\partial t} &= i\frac{\nu}{6\Delta x}\left[\frac{e^{i2\alpha\Delta x} - e^{-i2\alpha\Delta x}}{2i} - 8\frac{e^{i\alpha\Delta x} - e^{-i\alpha\Delta x}}{2i}\right]\hat{\boldsymbol{u}}^M \\ &\quad + \frac{\lambda_k}{6(\Delta x)^2}\left[-\frac{e^{i2\alpha\Delta x} + e^{-i2\alpha\Delta x}}{2} + 16\frac{e^{i\alpha\Delta x} + e^{-i\alpha\Delta x}}{2} - 15\right]\hat{\boldsymbol{u}}^M \\ &= \left[i\frac{\nu}{6\Delta x}\left[\sin(2\alpha\Delta x) - 8\sin(\alpha\Delta x)\right]\right. \\ &\quad \left. - \frac{\lambda_k}{3(\Delta x)^2}\left[\cos^2(2\alpha\Delta x) + 8(1 - \cos(\alpha\Delta x))\right]\right]\hat{\boldsymbol{u}}^M,\end{aligned} \tag{5.35}$$

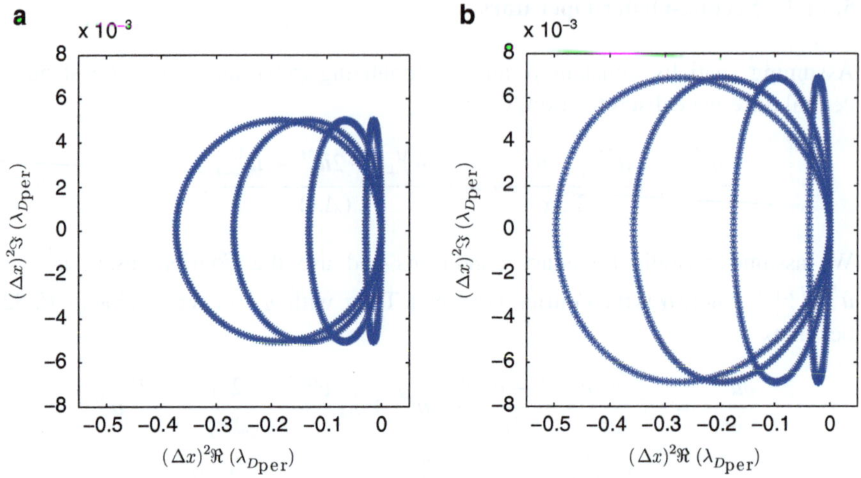

Fig. 5.5 Eigenvalues for order $M = 3$ Legendre polynomial chaos with 200 grid points, $\mu(\xi) \sim \mathscr{U}[0, 0.1]$, $\nu = 1$. (**a**) Second-order operators. (**b**) Fourth-order operators

which again is an ellipse in the negative half-plane. This is illustrated in Fig. 5.5, showing the eigenvalues of the second- and fourth-order periodic spatial discretization matrices $\boldsymbol{D}_{per}$. Since $\boldsymbol{D}_{per}$ is applied to periodic functions, no special boundary treatment is needed. Therefore, the entries of $\boldsymbol{D}_{per}$ are completely determined by the first and second derivative approximations of (5.32) and (5.34), respectively. In Fig. 5.5, the real part of the eigenvalues is denoted by $\Re$, and the complex part by $\Im$. Each of the eigenvalues $(\lambda_{\boldsymbol{B}})_k$, $k = 0, 1, 2, 3$, of $\boldsymbol{B}$ corresponds to one of the ellipses. For uniformly distributed μ, the range of the eigenvalues is bounded, and increasing the order of gPC does not increase the maximal eigenvalue significantly. Therefore, the order of gPC expansion has a negligible impact on the time-step restriction in this case.

For numerical stability, it is essential that the eigenvalues all be located in the negative half-plane. In the next section, we perform stability analysis for the more general case of an initial boundary value problem (with non-periodic boundary conditions).

5.5.2 *The Initial Boundary Value Problem*

In order to obtain stability of the semidiscretized problem for various orders of accuracy and non-periodic boundary conditions, we use discrete operators satisfying a summation-by-parts (SBP) property [12]. The SBP operators were introduced in Sect. 4.2.2, but for clarity we repeat some of the theory below.

Boundary conditions are imposed weakly through penalty terms, where the penalty parameters are chosen such that the numerical method is stable. Operators of order $2n$, $n \in \mathbb{N}$, in the interior of the domain are combined with boundary closures of order of accuracy n. For the advection-diffusion equation (5.1), this leads to the global order of accuracy $\min(n+2, 2n)$. We refer to [24] for a derivation of this result on accuracy.

As described above the first derivative operator is $u_x \approx \boldsymbol{P}^{-1}\boldsymbol{Q}\vec{u}$, where subscript x denotes partial derivative and $\boldsymbol{Q}$ satisfies (4.9), i.e.,

$$\boldsymbol{Q} + \boldsymbol{Q}^T = \mathrm{diag}(-1, 0, \ldots, 0, 1) \equiv \tilde{\boldsymbol{B}}. \tag{5.36}$$

The matrix $\boldsymbol{P}$ is symmetric and positive definite. For proof of stability of spatially varying viscosity $\mu(x, \xi)$, $\boldsymbol{P}$ must be diagonal, so we will only use SBP operators leading to a diagonal $\boldsymbol{P}$ norm.

To approximate the second derivative, we can use either the first derivative operator twice, or $\vec{\boldsymbol{u}}_{xx} \approx \boldsymbol{P}^{-1}(-\boldsymbol{M} + \tilde{\boldsymbol{B}}\boldsymbol{D})\vec{\boldsymbol{u}}$, where $\boldsymbol{M} + \boldsymbol{M}^T \geq 0$, $\tilde{\boldsymbol{B}}$ is given by (5.36), and $\boldsymbol{D}$ is a first-derivative approximation at the boundaries, with entries as given in 4.10.

Data on the boundaries are imposed weakly through a Simultaneous Approximation Term (SAT), introduced in [2]. Let the matrices $\boldsymbol{E}_1 = \mathrm{diag}(1, 0, \ldots, 0)$, $\boldsymbol{E}_m = \mathrm{diag}(0, \ldots, 0, 1)$ be used to position the boundary conditions, and let $\boldsymbol{\Sigma}_1^I$, $\boldsymbol{\Sigma}_1^V$ and $\boldsymbol{\Sigma}_m^V$ be penalty matrices to be chosen for stability. Let $\otimes$ denote the Kronecker product of two matrices $\boldsymbol{B}$ and $\boldsymbol{C}$ by

$$\boldsymbol{B} \otimes \boldsymbol{C} = \begin{bmatrix} [\boldsymbol{B}]_{11}\boldsymbol{C} & \ldots & [\boldsymbol{B}]_{1n}\boldsymbol{C} \\ \vdots & \ddots & \vdots \\ [\boldsymbol{B}]_{m1}\boldsymbol{C} & \ldots & [\boldsymbol{B}]_{mn}\boldsymbol{C} \end{bmatrix}.$$

The system (5.6) is discretized in space using SBP operators with the properties described above. For the general case of spatially varying viscosity $\mu(x, \xi)$, first-derivative operators will be successively applied to the viscosity term. An alternative, not considered here, is to use the compact SBP operators for $\partial/\partial x(b(x)\partial/\partial x)$ with $b(x) > 0$, developed in [14]. These operators have minimal stencil width for the order of accuracy. First, stability analysis for the general case of spatially varying viscosity is presented. As a further illustration of the SBP-SAT framework, then in the special case of spatially constant viscosity using compact second-derivative SBP operators stability analysis is also presented.

5.5.2.1 Spatially Varying Viscosity

Consider the case of a spatially varying $\mu = \mu(x, \xi)$, given by (5.6). Since μ depends on x, we cannot write the semidiscretized version of $\boldsymbol{B}$ as a Kronecker product. Instead, we introduce the block diagonal matrix

$$\hat{\boldsymbol{B}} = \mathrm{diag}(\boldsymbol{B}(x_1), \boldsymbol{B}(x_2), \ldots, \boldsymbol{B}(x_m)).$$

Note that $\hat{\boldsymbol{B}}$ and the matrix $(\boldsymbol{P}^{-1} \otimes \boldsymbol{I})$ commute, i.e.,

$$(\boldsymbol{P}^{-1} \otimes \boldsymbol{I})\hat{\boldsymbol{B}} = \hat{\boldsymbol{B}}(\boldsymbol{P}^{-1} \otimes \boldsymbol{I}). \tag{5.37}$$

Additionally, $\hat{\boldsymbol{B}}$ is symmetric, positive definite, and block diagonal. The matrix $(\boldsymbol{P}^{-1} \otimes \boldsymbol{I})\hat{\boldsymbol{B}}$ is a scaling of each diagonal block $\boldsymbol{B}(x_j)$ of $\hat{\boldsymbol{B}}$ with the factor $p_{jj}^{-1} > 0$. Thus, $(\boldsymbol{P}^{-1} \otimes \boldsymbol{I})\hat{\boldsymbol{B}}$ is symmetric and positive definite. The numerical approximation of (5.6) using SBP operators is given by

$$\begin{aligned}
&\frac{\partial \vec{\boldsymbol{u}}}{\partial t} + (\boldsymbol{P}^{-1}\boldsymbol{Q} \otimes \boldsymbol{V})\vec{\boldsymbol{u}} = (\boldsymbol{P}^{-1}\boldsymbol{Q} \otimes \boldsymbol{I})\hat{\boldsymbol{B}}(\boldsymbol{P}^{-1}\boldsymbol{Q} \otimes \boldsymbol{I})\vec{\boldsymbol{u}} \\
&+(\boldsymbol{P}^{-1} \otimes \boldsymbol{I})(\boldsymbol{E}_1 \otimes \boldsymbol{\Sigma}_1^I)(\vec{\boldsymbol{u}} - \boldsymbol{0}) + (\boldsymbol{P}^{-1} \otimes \boldsymbol{I})(\boldsymbol{Q}^T\boldsymbol{P}^{-1} \otimes \boldsymbol{I}) \\
&\times(\boldsymbol{E}_1 \otimes \boldsymbol{\Sigma}_1^V)(\vec{\boldsymbol{u}} - \boldsymbol{0}) + (\boldsymbol{P}^{-1} \otimes \boldsymbol{I})(\boldsymbol{E}_m \otimes \boldsymbol{\Sigma}_m^V)((\boldsymbol{P}^{-1}\boldsymbol{Q} \otimes \boldsymbol{I})\vec{\boldsymbol{u}} - \boldsymbol{0}),
\end{aligned} \tag{5.38}$$

where the first line corresponds to the discretization of the PDE, and the second and third lines enforce the homogeneous boundary conditions weakly, here expressed as $(\vec{\boldsymbol{u}} - \boldsymbol{0})$. Although the numerical experiments are performed with nonzero boundary conditions, it is sufficient to consider the homogeneous case in the analysis of stability.

Proposition 5.3. *The scheme in (5.38) with* $\boldsymbol{\Sigma}_m^V = -\boldsymbol{B}(x_m)$, $\boldsymbol{\Sigma}_1^V = \boldsymbol{B}(x_1)$, *and* $\boldsymbol{\Sigma}_1^I \le -\boldsymbol{V}/2$ *is stable in the sense of Definition 1.3.*

Proof. Multiplying (5.38) by $\vec{\boldsymbol{u}}^T(\boldsymbol{P} \otimes \boldsymbol{I})$ and replacing $\boldsymbol{Q} = \boldsymbol{E}_m - \boldsymbol{E}_1 - \boldsymbol{Q}^T$ in the first term of the right-hand side, we obtain

$$\begin{aligned}
&\vec{\boldsymbol{u}}^T(\boldsymbol{P} \otimes \boldsymbol{I})\frac{\partial \vec{\boldsymbol{u}}}{\partial t} + \overbrace{\vec{\boldsymbol{u}}^T(\boldsymbol{Q} \otimes \boldsymbol{V})\vec{\boldsymbol{u}}}^{\text{Advection term}} = \overbrace{\vec{\boldsymbol{u}}^T(\boldsymbol{E}_m \otimes \boldsymbol{I})\hat{\boldsymbol{B}}(\boldsymbol{P}^{-1}\boldsymbol{Q} \otimes \boldsymbol{I})\vec{\boldsymbol{u}}}^{\text{Viscous terms}} \\
&\underbrace{-\vec{\boldsymbol{u}}^T(\boldsymbol{E}_1 \otimes \boldsymbol{I})\hat{\boldsymbol{B}}(\boldsymbol{P}^{-1}\boldsymbol{Q} \otimes \boldsymbol{I})\vec{\boldsymbol{u}} - \vec{\boldsymbol{u}}^T(\boldsymbol{Q}^T \otimes \boldsymbol{I})\hat{\boldsymbol{B}}(\boldsymbol{P}^{-1} \otimes \boldsymbol{I})(\boldsymbol{Q} \otimes \boldsymbol{I})\vec{\boldsymbol{u}}}_{\text{Viscous terms}} \\
&+\underbrace{\vec{\boldsymbol{u}}^T(\boldsymbol{E}_1 \otimes \boldsymbol{\Sigma}_1^I)\vec{\boldsymbol{u}}}_{\text{Adv. penalty term}} + \underbrace{\vec{\boldsymbol{u}}^T(\boldsymbol{Q}^T\boldsymbol{P}^{-1} \otimes \boldsymbol{I})(\boldsymbol{E}_1 \otimes \boldsymbol{\Sigma}_1^V)\vec{\boldsymbol{u}}}_{\text{Left viscous penalty term}} \\
&+\underbrace{\vec{\boldsymbol{u}}^T(\boldsymbol{E}_m \otimes \boldsymbol{\Sigma}_m^V)(\boldsymbol{P}^{-1}\boldsymbol{Q} \otimes \boldsymbol{I})\vec{\boldsymbol{u}}}_{\text{Right viscous penalty term}}.
\end{aligned} \tag{5.39}$$

The right viscous penalty term and the first viscous term cancel if we set $\boldsymbol{\Sigma}_m^V = -\boldsymbol{B}(x_m)$. Adding the transpose of the remaining terms of (5.39) to themselves and using (5.37), we arrive at the energy equation

$$\frac{\partial}{\partial t}\left\|\vec{u}\right\|^2_{P\otimes I} + \overbrace{\vec{u}^T(E_m\otimes V)\vec{u} - \vec{u}^T(E_1\otimes V)\vec{u}}^{\text{Advection boundary terms}} =$$
$$= \underbrace{-\vec{u}^T(E_1\otimes I)\hat{B}(P^{-1}Q\otimes I)\vec{u} - \vec{u}^T(Q^TP^{-1}\otimes I)\hat{B}(E_1\otimes I)\vec{u}}_{\text{Viscous terms}}$$
$$-2\vec{u}^T(Q^T\otimes I)\hat{B}(P^{-1}\otimes I)(Q\otimes I)\vec{u} + \underbrace{2\vec{u}^T(E_1\otimes \Sigma_1^I)\vec{u}}_{\text{Adv. penalty term}}$$
$$+\underbrace{\vec{u}^T(Q^TP^{-1}\otimes I)(E_1\otimes\Sigma_1^V)\vec{u} + \vec{u}^T(E_1\otimes\Sigma_1^V)(P^{-1}Q\otimes I)\vec{u}}_{\text{Left viscous penalty terms}}. \tag{5.40}$$

The viscous terms from the PDE and the left viscous penalty terms cancel if we set $\Sigma_1^V = B(x_1)$. Pairing the second advective boundary term with the advective penalty term for stability and choosing $\Sigma_1^I = -\delta V$, where $\delta \in \mathbb{R}$, leads to

$$\frac{\partial}{\partial t}\left\|\vec{u}\right\|^2_{P\otimes I} = u_1^T(1-2\delta)vu_1 - u_m^Tvu_m - 2\left[(Q\otimes I)\vec{u}\right]^T\hat{B}(P^{-1}\otimes I)\left[(Q\otimes I)\vec{u}\right]. \tag{5.41}$$

For $\delta \geq 1/2$, i.e., $\Sigma_1^I \leq -V/2$, the energy rate (5.41) shows that the scheme (5.38) with variable B is stable as defined in Definition 1.3, as the norm of $\vec{u}$ decays with time.

5.5.2.2 Spatially Constant Viscosity

For the case of spatially constant viscosity $\mu(\xi)$, we use compact second-derivative SBP operators. We show that the choice of penalty matrices is similar to the case of spatially varying viscosity $\mu(x,\xi)$ presented in the preceding section. The scheme is given by

$$\frac{\partial\vec{u}}{\partial t} + (P^{-1}Q\otimes V)\vec{u} = (P^{-1}(-M+\tilde{B}D)\otimes B)\vec{u}$$
$$+(P^{-1}\otimes I)(E_1\otimes\Sigma_1^I)(\vec{u}-\mathbf{0}) + (P^{-1}\otimes I)(D^T\otimes I)(E_1\otimes\Sigma_1^V)(\vec{u}-\mathbf{0})$$
$$+(P^{-1}\otimes I)(E_m\otimes\Sigma_m^V)((D\otimes I)\vec{u}-\mathbf{0}). \tag{5.42}$$

Proposition 5.4. *The scheme in (5.42) with the parameters* $\Sigma_1^V = B$, $\Sigma_m^V = -B$, *and* $\Sigma_1^I \leq -V/2$ *is stable in the sense of Definition 1.3.*

Proof. Multiplying (5.42) by $\vec{u}^T(P\otimes I)$ and using $\tilde{B} = -E_1 + E_m$, we get

$$\vec{u}^T(\boldsymbol{P}\otimes\boldsymbol{I})\frac{\partial\vec{u}}{\partial t}+\overbrace{\vec{u}^T(\boldsymbol{Q}\otimes\boldsymbol{V})\vec{u}}^{\text{Advection term}}=$$

$$=\overbrace{-\vec{u}^T(\boldsymbol{M}\otimes\boldsymbol{B})\vec{u}-\vec{u}^T(\boldsymbol{E}_1\boldsymbol{D}\otimes\boldsymbol{B})\vec{u}+\vec{u}^T(\boldsymbol{E}_m\boldsymbol{D}\otimes\boldsymbol{B})\vec{u}}^{\text{Viscous terms}}+\overbrace{\vec{u}^T(\boldsymbol{E}_1\otimes\boldsymbol{\Sigma}_1^I)\vec{u}}^{\text{Inviscid penalty term}}$$

$$+\underbrace{\vec{u}^T(\boldsymbol{D}^T\otimes\boldsymbol{I})(\boldsymbol{E}_1\otimes\boldsymbol{\Sigma}_1^V)\vec{u}}_{\text{Left viscous penalty term}}+\underbrace{\vec{u}^T(\boldsymbol{E}_m\otimes\boldsymbol{\Sigma}_m^V)(\boldsymbol{D}\otimes\boldsymbol{I})\vec{u}}_{\text{Right viscous penalty term}}. \tag{5.43}$$

As in the case of variable viscosity, setting $\boldsymbol{\Sigma}_1^V = \boldsymbol{B}$ and $\boldsymbol{\Sigma}_m^V = -\boldsymbol{B}$ cancels the viscous boundary terms of the ODE. Using the relation

$$\vec{u}^T(\boldsymbol{Q}\otimes\boldsymbol{V})\vec{u}=\vec{u}^T\left(\frac{1}{2}(\boldsymbol{Q}+\boldsymbol{Q}^T)\otimes\boldsymbol{V}\right)\vec{u}+\underbrace{\vec{u}^T\left(\frac{1}{2}(\boldsymbol{Q}-\boldsymbol{Q}^T)\otimes\boldsymbol{V}\right)\vec{u}}_{=0}$$

$$=\frac{1}{2}\vec{u}^T((-\boldsymbol{E}_1+\boldsymbol{E}_m)\otimes\boldsymbol{V})\vec{u}, \tag{5.44}$$

we arrive at

$$\vec{u}^T(\boldsymbol{P}\otimes\boldsymbol{I})\frac{\partial\vec{u}}{\partial t}=-\vec{u}^T(\boldsymbol{M}\otimes\boldsymbol{B})\vec{u}+\vec{u}^T(\boldsymbol{E}_1\otimes(\boldsymbol{V}/2+\boldsymbol{\Sigma}_1^I))\vec{u}-\vec{u}^T(\boldsymbol{E}_m\otimes\boldsymbol{V}/2)\vec{u} \tag{5.45}$$

Finally, setting $\boldsymbol{\Sigma}_1^I = -\delta\boldsymbol{V}$ as in Sect. 5.5.2.1 and adding the transpose of (5.45) to itself, we get the energy estimate

$$\frac{\partial}{\partial t}\|\vec{u}\|^2_{(\boldsymbol{P}\otimes\boldsymbol{I})}=\boldsymbol{u}_1^T(1-2\delta)\boldsymbol{v}\boldsymbol{u}_1-\boldsymbol{u}_m^T\boldsymbol{v}\boldsymbol{u}_m-\vec{u}^T((\boldsymbol{M}+\boldsymbol{M}^T)\otimes\boldsymbol{B})\vec{u}. \tag{5.46}$$

Since $\boldsymbol{M}+\boldsymbol{M}^T$ and $\boldsymbol{B}$ are positive definite, the relation (5.46) with $\delta \geq 1/2$, i.e., $\boldsymbol{\Sigma}_1^I \leq -\boldsymbol{V}/2$, proves that the scheme in (5.42) is stable since it satisfies the conditions of Definition 1.3.

5.5.3 Eigenvalues of the Total System Matrix

The semidiscrete scheme (5.42) is an ODE system of the form

$$\frac{\partial\vec{u}}{\partial t}=\boldsymbol{D}_{tot}\vec{u},$$

whose properties are determined by the complex-valued eigenvalues of the total system matrix $\boldsymbol{D}_{tot}$. The eigenvalues of $\boldsymbol{D}_{tot}$ must all have negative real parts for stability. The utmost right-lying eigenvalue determines the slowest decay rate, and thus the speed of convergence to steady-state (see [16,17]). The total spatial operator defined by the scheme (5.42) with $\boldsymbol{\Sigma}_1^V = \boldsymbol{B}$, $\boldsymbol{\Sigma}_m^V = -\boldsymbol{B}$, and $\boldsymbol{\Sigma}_1^I = -\boldsymbol{V}/2$ is given by the matrix

$$\boldsymbol{D}_{tot} = (\boldsymbol{P}^{-1} \otimes \boldsymbol{I})\left(-(\boldsymbol{Q} + \boldsymbol{E}_1/2) \otimes \boldsymbol{V} + (\boldsymbol{D}^T\boldsymbol{E}_1 - \boldsymbol{E}_1\boldsymbol{D} - \boldsymbol{M}) \otimes \boldsymbol{B}\right). \tag{5.47}$$

The location in the complex plane of the eigenvalues of $\boldsymbol{D}_{tot}$ depends on the distribution of μ, the spatial step Δx, and the ratio between viscosity and advective speed.

Figure 5.6 depicts the eigenvalues of $\boldsymbol{D}_{tot}$ for uniform $\mu(\xi) \sim \mathscr{U}[0, 0.04]$, $v = 1$, different orders of polynomial chaos, and number of spatial grid points. The fourth-

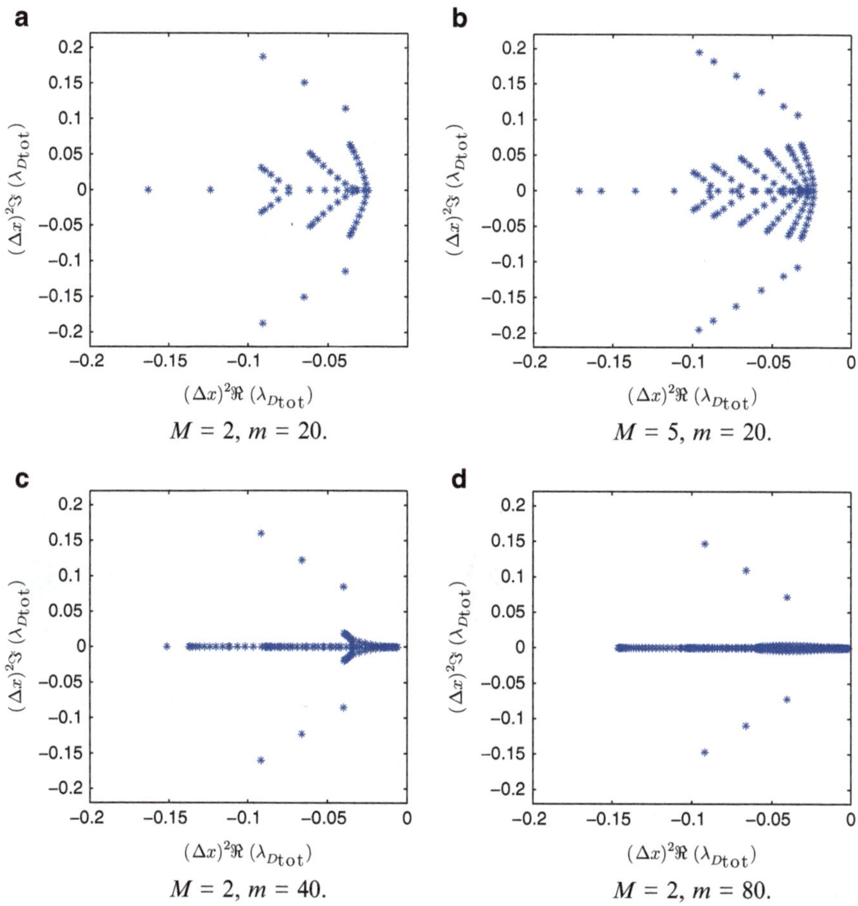

Fig. 5.6 Eigenvalues of the total operator $\boldsymbol{D}_{tot}$ (including penalty terms). Comparison of different orders of gPC (**a**) and (**b**), and different grid sizes (**a**), (**c**), and (**d**)

order SBP operators have been used, and penalty coefficients are chosen according to the stability analysis above. The eigenvalues all have negative real parts, showing that the discretizations are indeed stable. Note that for an order of gPC expansion M, there will be $M + 1$ eigenvalues for each one eigenvalue of the corresponding deterministic system matrix. The groups of $M + 1$ eigenvalues are clustered around the corresponding eigenvalue of the deterministic system matrix. When the range of possible viscosity values (uncertainty) is increased, the spreading of the eigenvalues within each cluster increases. When the mean of the viscosity is increased, the eigenvalues with a nonzero complex part will move farther away from the origin.

The change of location of the eigenvalues with increasing order of gPC expansion gives an idea how the time-step restriction changes. Figure 5.7 shows the eigenvalues of the total system matrix for uniform and lognormal μ for first-order (left) and fourth-order (right) gPC. For the random viscosities to be comparable, the coefficients are chosen such that the first and second moments of the uniform and the lognormal μ match each other. For low-order polynomial chaos expansions, the eigenvalues are close to each other and the systems are similar in terms of stiffness. As the order of gPC expansion is increased, the scattering of the eigenvalues of $\boldsymbol{B}$ resulting from lognormal μ increases (μ is unbounded). Hence, the stochastic Galerkin system becomes stiffer with increasing order of gPC. The time-step restriction for the uniform viscosity does not change significantly with the order of gPC. The fourth-order operators are a factor of approximately 1.5 stiffer than the second-order operators. Here, we calculate stiffness as

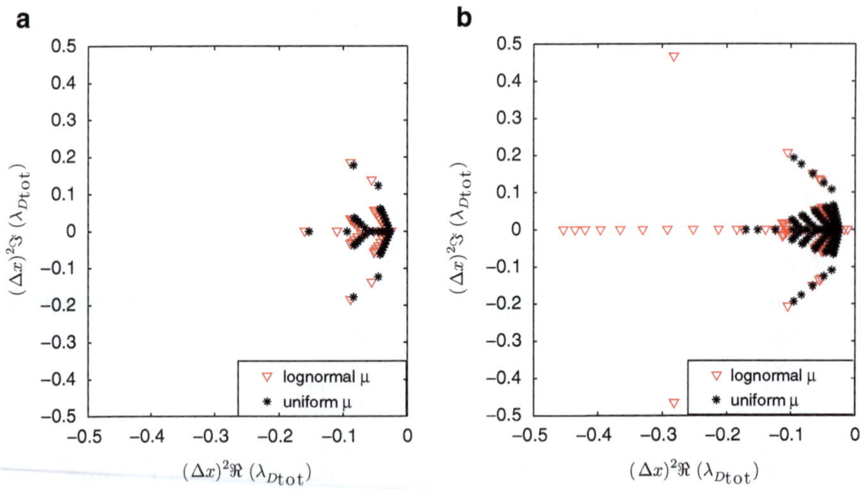

Fig. 5.7 Eigenvalues of the total operator for $m = 20$ and different orders of gPC. Here the viscosity μ has mean $\langle\mu\rangle = 0.02$ and variance $\mathrm{Var}(\mu) = 3.33 \times 10^{-5}$, and has uniform and lognormal distributions. **(a)** $M = 1$. **(b)** $M = 4$

$$\rho_{\text{stiff}} = \frac{\max|\lambda_{\boldsymbol{D}_{tot}}|}{\min|\lambda_{\boldsymbol{D}_{tot}}|},$$

where $|\lambda_{\boldsymbol{D}_{tot}}|$ denotes the absolute values of the complex eigenvalues of the total spatial operator $\boldsymbol{D}_{tot}$.

5.5.4 Convergence to Steady-State

As we let $t \to \infty$, the problem (5.6) with $\boldsymbol{B}(x) > 0$ will reach steady-state, i.e., it will satisfy $\partial \boldsymbol{u}^M/\partial t = \boldsymbol{0}$. This situation can be formulated as a time-independent problem with solution $\tilde{\boldsymbol{u}^M}$, that satisfies

$$\begin{aligned} \boldsymbol{V}\frac{\partial \tilde{\boldsymbol{u}}}{\partial x} &= \frac{\partial}{\partial x}\left(\boldsymbol{B}(x)\frac{\partial \tilde{\boldsymbol{u}}}{\partial x}\right), \\ \tilde{\boldsymbol{u}}(x=0) &= \tilde{\boldsymbol{g}}_0, \\ \frac{\partial \tilde{\boldsymbol{u}}(x)}{\partial x}|_{x=1} &= \tilde{\boldsymbol{g}}_1. \end{aligned} \tag{5.48}$$

By subtracting (5.48) from (5.6), we get the initial boundary value problem for the deviation $\boldsymbol{e} = \boldsymbol{u} - \tilde{\boldsymbol{u}}$ from steady-state,

$$\frac{\partial \boldsymbol{e}}{\partial t} + \boldsymbol{V}\frac{\partial \boldsymbol{e}}{\partial x} = \frac{\partial}{\partial x}\left(\boldsymbol{B}(x)\frac{\partial \boldsymbol{e}}{\partial x}\right), \tag{5.49}$$

$$\begin{aligned} \boldsymbol{e}(0,t) &= \boldsymbol{0}, \\ \frac{\partial \boldsymbol{e}(x,t)}{\partial x}|_{x=1} &= \boldsymbol{0}, \\ \boldsymbol{e}(x,0) &= \boldsymbol{u}_{init}(x) - \tilde{\boldsymbol{u}}(x) = \boldsymbol{e}_0(x), \end{aligned} \tag{5.50}$$

where it has been used that as $t \to \infty$, the boundary data must be independent of time and vanish. The problem (5.49) can be semidiscretized analogously to the numerical schemes presented in Sect. 5.5.2. Thus, with $\boldsymbol{D}_{tot}$ defined in (5.47), the aim is to solve the initial value problem

$$\frac{\partial \vec{\boldsymbol{e}}}{\partial t} = \boldsymbol{D}_{tot}\vec{\boldsymbol{e}}, \quad t > 0, \tag{5.51}$$

$$\vec{\boldsymbol{e}} = \vec{\boldsymbol{e}}_0(x), \quad t = 0, \tag{5.52}$$

with the solution $\vec{\boldsymbol{e}}(x,t) = \vec{\boldsymbol{e}}_0(x)\exp(\boldsymbol{D}_{tot}t)$. The largest real component of the eigenvalues of $\boldsymbol{D}_{tot}$, denoted by $\max \Re(\lambda_{\boldsymbol{D}_{tot}})$, must be negative; otherwise, the solution will not converge to steady-state. The more negative $\max \Re(\lambda_{\boldsymbol{D}_{tot}})$ is, the faster the convergence to steady-state.

Although the boundary conditions may be altered in different ways to accelerate the convergence to steady-state [17], we use the weak imposition of boundary conditions described in Sect. 5.5.2 and compare the convergence to steady-state for a diagonalizable stochastic Galerkin system with that of the stochastic collocation method. The number of iterations to reach convergence to steady-state depends on the size of the time-step and the exponential decay of the solution, governed by the rightmost lying eigenvalue of the total system matrix, $\max \Re(\lambda_{\boldsymbol{D}_{tot}})$. For each stochastic quadrature point of the advection-diffusion equation, there is a maximal time-step as well as a maximal eigenvalue of the total system matrix. For stochastic Galerkin, each scalar instance of the advection-diffusion equation corresponds to one of the eigenvalues of $\boldsymbol{B}$, and for stochastic collocation, each instance corresponds to μ evaluated at a stochastic quadrature point.

Explicit time integration together with various convergence acceleration techniques such as residual smoothing, local time-stepping and multigrids are the most common methods for reaching steady-state in flow calculations [9–11]. In this simplified case, explicit time integration with the maximum possible time-step possible illustrates this scenario.

Figure 5.8 depicts the maximum time-step and the maximum eigenvalue of $\boldsymbol{D}_{tot}$ for each one instance of an advection-diffusion equation for different approximation orders of the gPC and stochastic collocation. If diagonalization is possible, and for sufficiently high orders of stochastic Galerkin, the scalar instances of the continuous advection-diffusion equation with the most negative $\max \Re(\lambda_{\boldsymbol{D}_{tot}})$ converge to steady-state faster than the corresponding instances of stochastic collocation. However, the severe time-step limit of stochastic Galerkin implies that a large number of time-steps is needed to reach steady-state numerically with explicit time-stepping. It is not clear from Fig. 5.8 alone whether stochastic Galerkin or stochastic collocation reaches steady-state numerically in the smaller number of time-steps. This uncertainty will be investigated in Sect. 5.6.2.

For non-diagonalizable stochastic Galerkin, the local bounds of Fig. 5.8 on time-steps and maximum eigenvalues no longer apply. Instead, the most severe local time-step limit and eigenvalue will dominate the entire stochastic Galerkin system, with deteriorating performance as a consequence. Stochastic collocation is still subject to local time-step restrictions and local maximum eigenvalues, and is expected to converge faster to steady-state than stochastic Galerkin.

A practical algorithm for steady-state calculations should be designed to be as efficient as possible in terms of computational cost. For instance, one may use an implicit/explicit scheme as devised in [30] for stochastic diffusion problems. What we presented above is not an efficient algorithm for steady-state calculations; rather it is an analysis of the properties of the semidiscrete system leading to convergence to steady-state.

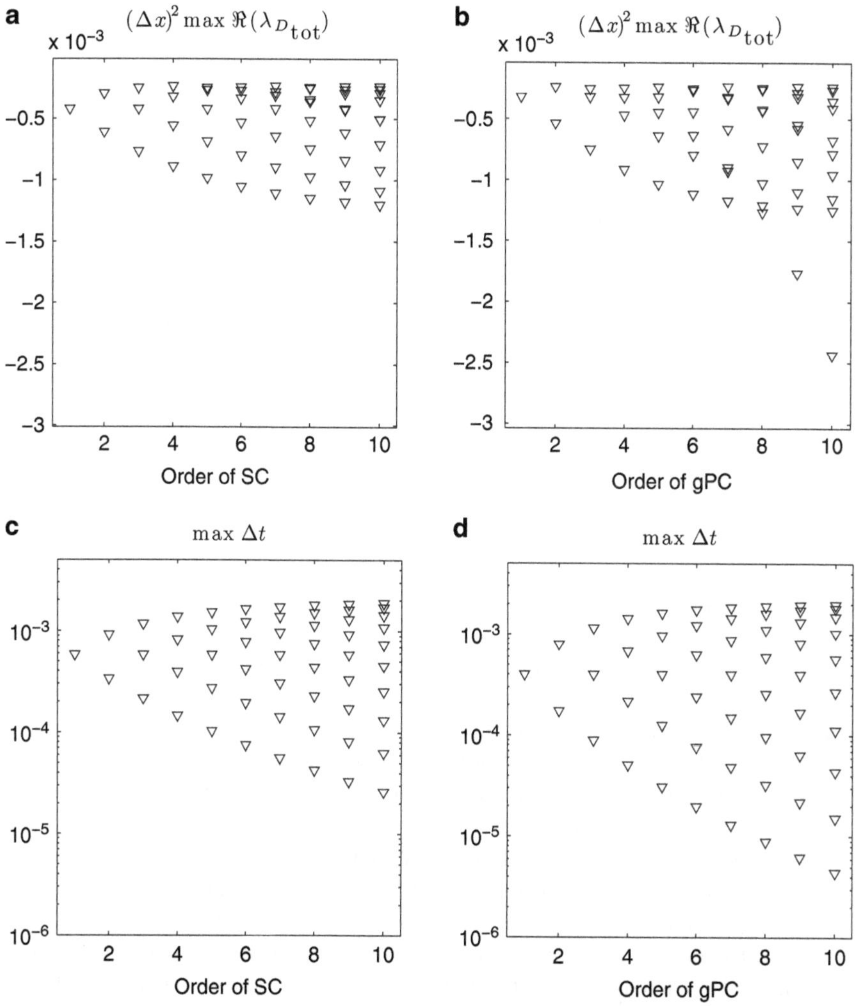

Fig. 5.8 Convergence to steady-state depends on the limit on Δt and $\max \Re\left(\lambda_{D_{tot}}\right)$. These quantities are plotted for lognormal viscosity $\mu(\xi) = 0.02+0.05\exp(\xi)$, $\xi \sim \mathcal{N}(0,1)$. Stochastic collocation (*left*) and stochastic Galerkin (*right*). (**a**) $\max \Re(\lambda_{D_{tot}})$ for each quadrature point as a function of the order of stochastic collocation. (**b**) $\max \Re(\lambda_{D_{tot}})$ for the scalar advection-diffusion equations (one for each eigenvalue (λ_B)) for different orders of stochastic Galerkin. (**c**) Time-step limit for each quadrature point of stochastic collocation. (**d**) Time-step limit for each scalar advection diffusion equation (diagonalizable system) of the stochastic Galerkin method

5.6 Numerical Results

In the numerical examples of this Section, we use a fourth-order Runge-Kutta method for the time integration and the fourth-order accurate SBP-SAT scheme in space. The matrix operators can be found in [15]. The scalar problem (5.1) with spatially independent μ is solved for the initial function

$$u_0(x,\xi) = \frac{\rho_0}{\sqrt{4\pi\mu(\xi)\tau}} \exp\left(-\frac{(x-(x_0+v\tau))^2}{4\mu(\xi)\tau}\right), \quad \rho_0 > 0,\ x_0 \in [0,1],\ \tau > 0,$$

for which the analytical solution at time t is given by

$$u(x,t,\xi) = \frac{\rho_0}{\sqrt{4\pi\mu(\xi)(t+\tau)}} \exp\left(-\frac{(x-(x_0+v(t+\tau)))^2}{4\mu(\xi)(t+\tau)}\right). \tag{5.53}$$

For the spatially varying $\mu(x,\xi)$, we employ the method of manufactured solutions [20,21] where we get the same solution as in the case of spatially constant $\mu(\xi)$ with the aid of an appropriate source function $s(x,t,\xi)$ in (5.1). The source function is given by

$$\begin{aligned} s(x,t,\xi) \\ = \frac{(x-(x_0+v(t+\tau)))(2\mu(x,\xi)-\mu_x(x,\xi)(x-(x_0+v(t+\tau))))\mu_x(x,\xi)u}{4\mu^2(x,\xi)(t+\tau)} \\ + \frac{\mu_x^2(x,t)}{2\mu(x,t)}u. \end{aligned} \tag{5.54}$$

The stochastic reference solution (5.53) is projected onto the gPC basis functions using a high-order numerical quadrature. The order N of the quadrature is chosen sufficiently large so that the difference between two successive reference solutions of order $N-1$ and N are several orders of magnitude smaller than the difference between the solution from the numerical scheme and the reference solution.

Figure 5.9 illustrates the convergence as the spatial grid is refined for constant order of gPC, $M = 12$ and $N = 13$ collocation points. For this high-order stochastic representation, the theoretical fourth-order convergence rate is attained for the mean using stochastic Galerkin and stochastic collocation. For the variance, the stochastic truncation error becomes visible for fine spatial meshes with lognormal μ (see Fig. 5.9b). There is no significant difference in performance between stochastic collocation and stochastic Galerkin for this test case.

5.6.1 The Inviscid Limit

The theoretical results for the advection-diffusion problem are based on $\mu > 0$. When μ is arbitrarily close to 0 (but non-negative), the problem becomes nearly hyperbolic. In the stochastic setting, this happens with nonzero probability whenever $\mu(\xi) \in [0,c]$, $c > 0$. For small μ the mesh must be very fine, otherwise the mesh Reynolds number requirement discussed in Sect. 5.4 will be violated. This is illustrated in Fig. 5.10 (numerical solution left and error right) for results obtained with fourth-order SBP operators, $v = 1$ and $\mu \sim \mathscr{U}[0.01, 0.19]$. Note that the error is maximal close to the inviscid limit of $\mu = 0.01$. The solution (5.53) is a Gaussian

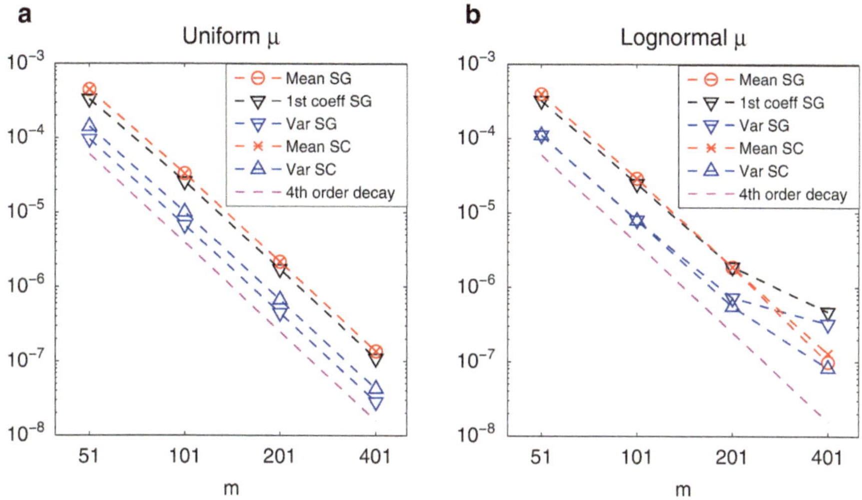

Fig. 5.9 Convergence with respect to the spatial discretization using stochastic Galerkin (*SG*) and stochastic collocation (*SC*). Plotted are norms of the absolute errors in mean, first coefficient and variance with $M = 12$ order of generalized Legendre/Hermite chaos, and $N = 13$ quadrature points for stochastic collocation. (**a**) Uniform $\mu \in (0.05, 0.15)$. (**b**) Lognormal $\mu = 0.05 + 0.05 \exp(\xi)$

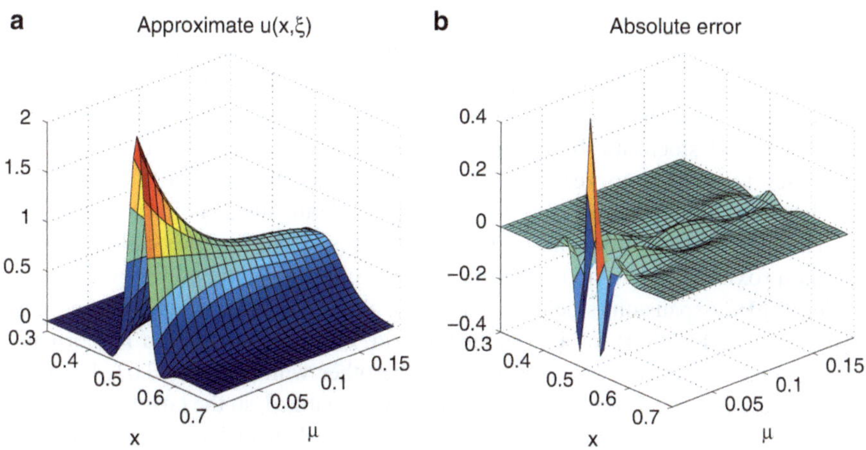

Fig. 5.10 Approximate solution with $M = 3$ order of Legendre chaos. (**a**) Solution at $t = 0.005$, $\Delta x = 0.002$. (**b**) Error of the approximate solution

in space for any fixed value of ξ and t and varies exponentially in x with the inverse of μ. Thus, spatial convergence requires a fine mesh for small μ. Deterioration of the convergence properties for small μ is a well-known phenomenon for other problems with a parabolic term, e.g., the Navier-Stokes equations in the inviscid limit.

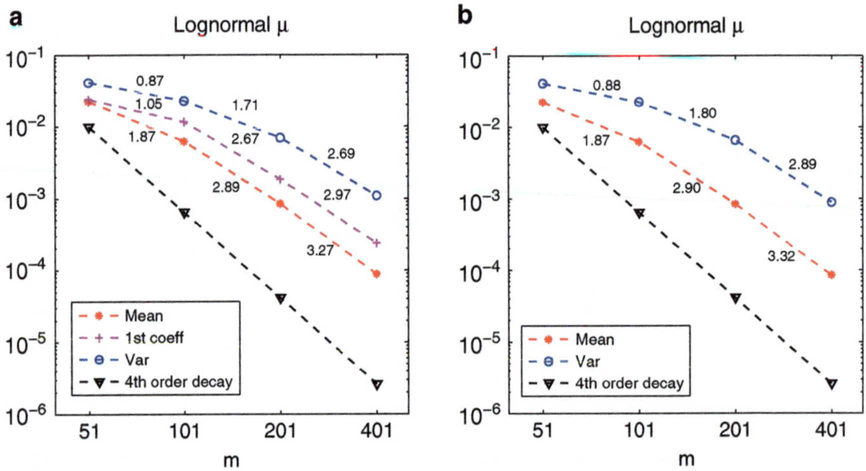

Fig. 5.11 Spatial convergence, lognormal viscosity, $\mu = 0.02\exp(\xi)$. $T = 0.001$ and $\tau = 0.005$. Local numerical order of convergence indicated in the plots. (**a**) Stochastic Galerkin with $M = 9$. (**b**) Stochastic collocation with 10 quadrature points

Figure 5.11 shows the convergence in space for $\mu = 0.02\exp(\xi)$, $\xi \sim \mathcal{N}(0,1)$, using the stochastic Galerkin method (left) and the stochastic collocation method (right). When μ approaches zero, the gradients become steeper, which requires finer resolution. The fourth-order convergence rate is not obtained for these coarse meshes. As long as the stochastic basis is rich enough to represent the uncertainty, the choice of stochastic collocation versus stochastic Galerkin has no significant effect either on the rate of spatial convergence, or on the actual error. However, the number of stochastic basis functions needed for a certain level of resolution increases as μ goes to zero; therefore, a simultaneous increase in spatial and stochastic resolution is necessary for convergence in the inviscid limit.

The performance of stochastic Galerkin versus stochastic collocation depends on the proximity to the inviscid limit. Figure 5.12 shows the convergence in the order of gPC expansion (stochastic Galerkin) and the number of quadrature points (stochastic collocation) for a fixed spatial grid. Two cases of shifted lognormal μ are compared; one with $\mu_{min} = 0.2$ and the other with $\mu_{min} = 0.01$. For these cases, the stochastic Galerkin system can be diagonalized, so the cost for stochastic Galerkin with an expansion order $M - 1$ is equivalent to the cost of stochastic collocation with M quadrature points. If the problem is diffusion dominated, stochastic Galerkin is the more efficient method. If the viscosity is close to zero with some nonzero probability, the difference in performance decreases. Low viscosity sharpens the solution's features. The effect of this on the spatial convergence is seen in the low-viscosity case ($\mu_{min} = 0.01$) in Fig. 5.12b, where the spatial truncation error becomes visible for high-order polynomial chaos expansions. Due to the fixed number of spatial grid points, the convergence rate decreases for high-order stochastic representations. With a sufficiently fine mesh, it is possible to show exponential convergence rate for any given order of stochastic representation.

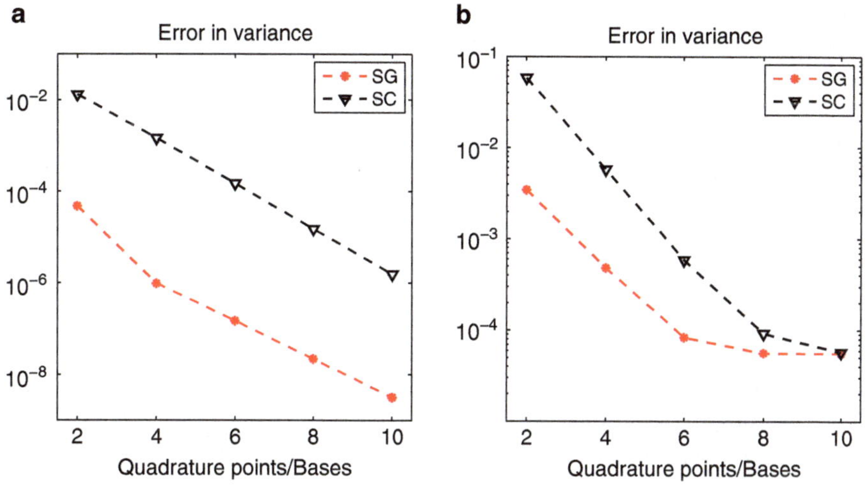

Fig. 5.12 Stochastic Galerkin (SG) and stochastic collocation (SC) as a function of the order of gPC/number of quadrature points. Fixed mesh of 201 spatial points. (**a**) Lognormal viscosity, $\mu = 0.2 + 0.01\exp(\xi)$. (**b**) Lognormal viscosity, $\mu = 0.01 + 0.01\exp(\xi)$

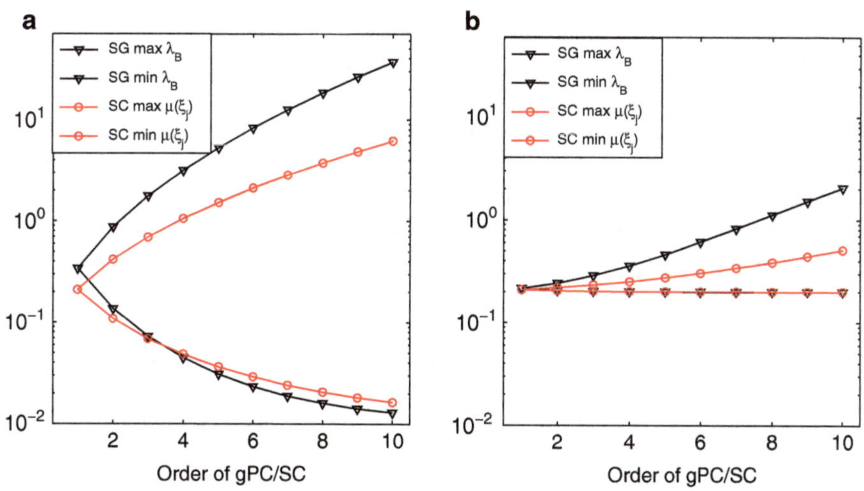

Fig. 5.13 Minimum and maximum viscosity for different orders of stochastic Galerkin (*SG*) and stochastic collocation (*SC*) for two different distributions of μ. This corresponds to the minimum and maximum λ_B for SG and to the minimum and maximum $\mu(\xi)$ for SC. (**a**) Lognormal viscosity, $\mu = 0.01 + 0.2\exp(\xi)$. (**b**) Lognormal viscosity, $\mu = 0.2 + 0.01\exp(\xi)$

Both the stochastic collocation and diagonalizable stochastic Galerkin rely on a set of scalar advection-diffusion problems, with the difference between the methods lying in the choice of stochastic viscosity point values and the postprocessing used to obtain statistics of interest. Figure 5.13 displays the difference in the range of the effective values of μ for stochastic collocation and stochastic Galerkin (this corresponds to the range of eigenvalues of $\boldsymbol{B}$ for stochastic Galerkin). From a purely numerical point of view, stochastic Galerkin poses an additional challenge compared to stochastic collocation in that a wider range of scales of diffusion must be handled simultaneously, as shown in Fig. 5.13. If we were to choose the eigenvalues of the matrix $\boldsymbol{B}$ as the collocation points, the two methods would differ only in the postprocessing.

5.6.2 Steady-State Calculations

Let the time of numerical convergence to steady-state be defined as the time T_{ss} when the discretized residual $\vec{\boldsymbol{e}}$ satisfies $\|\vec{\boldsymbol{e}}\|_{2,\Delta x} = \left(\Delta x \sum_{i=1}^{m}(\boldsymbol{e}(x_i, T_{ss}))^2\right)^{1/2} < tol$, where tol is a numerical tolerance to be chosen a priori. When μ is sufficiently large so that diffusion is the dominating feature compared to advection, larger values of the range of μ imply that T_{ss} decreases, and steady-state is reached sooner. The number of iterations to steady-state (i.e., the number of time-steps $T_{ss}/\Delta t$ for a uniform Δt) is inversely proportional to Δt. On the other hand, the limit on Δt decreases with μ. Hence, there is a trade-off in the number of iterations to steady-state between the size of time-step and the eigenvalues or quadrature point values of μ. In Fig. 5.14, this trade-off is explored for a shifted lognormal $\mu = c_1 + c_2 \exp(\xi)$ with different choices of c_1 and c_2.

From the previous analysis and Figs. 5.8 and 5.13, we have observed how the eigenvalues grow with the order M of gPC. For the most advection-dominated case, Fig. 5.14a, the number of iterations grows superlinearly with the number of quadrature points in the stochastic collocation approach. The same growth also occurs for up to order $M = 8$ of stochastic Galerkin. For this case, the

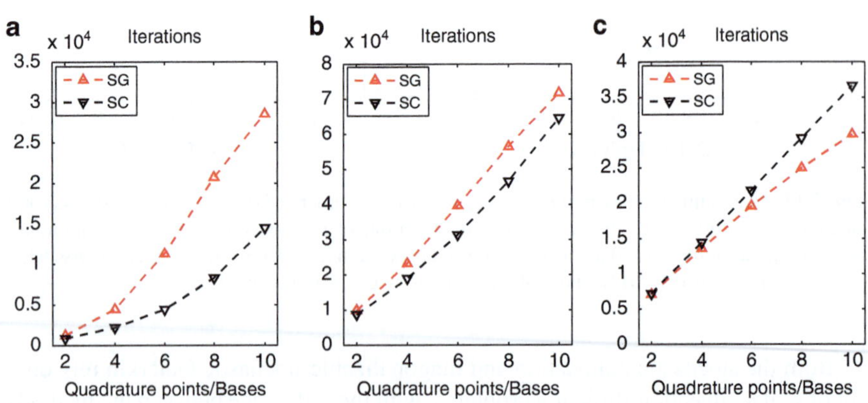

Fig. 5.14 Number of iterations to steady-state for different lognormal viscosity $\mu = c_1 + c_2 \exp(\xi)$ using stochastic Galerkin and stochastic collocation. Here $tol = 10^{-6}$. **(a)** $\mu = 0.02 + 0.005 \exp(\xi)$. **(b)** $\mu = 0.1 + 0.01 \exp(\xi)$. **(c)** $\mu = 2 + 0.2 \exp(\xi)$

fastest convergence to steady-state is obtained for stochastic collocation, where the range of possible μ values is narrower than in the case of stochastic Galerkin (see Fig. 5.13). In the more diffusive case, i.e., Fig. 5.14b, the relative speed-up for stochastic collocation versus stochastic Galerkin is less pronounced. In the most diffusive case considered here, Fig. 5.14c, the number of stochastic Galerkin iterations required to steady-state is a sublinear function of the order of gPC. In this case, the largest eigenvalues λ_B yield advection-diffusion equations that converge within a relatively short time T_{ss}, which compensates for a severe time-step restriction. For these diffusive cases, stochastic Galerkin is more efficient than stochastic collocation. The stochastic Galerkin problem has been diagonalized to make the computational cost per iteration similar. In summary, Fig. 5.14 shows that stochastic collocation converges faster than stochastic Galerkin to steady-state for problems that are advection dominated or moderately diffusive. For diffusion-dominated flows, stochastic Galerkin converges faster to steady-state than does stochastic collocation.

5.7 Summary and Conclusions

A stochasic Galerkin formulation of the advection-diffusion equation is a relatively simple linear problem. It provides a controlled, but sufficiently complex, setting for demonstration of numerical phenomena that need to be addressed in more complex linear and nonlinear problems. In this Chapter, summation-by-parts operators and weak boundary treatment have been applied to a stochastic Galerkin formulation of the advection-diffusion equation. We have presented conditions for monotonicity for stochastic, but homogeneous in space, viscosity, and stable schemes for the more general case of spatially varying and uncertain viscosity. Stochastic Galerkin projection should preserve well-posedness, as shown for the projection of the viscosity where we require the viscosity matrix B to be positive semidefinte. If B has a negative eigenvalue, the problem is ill-posed. A corresponding problem exists for non-intrusive methods, where the stochastic quadrature or collocation points must be chosen such that the viscosity remains non-negative for all evaluations of the PDE.

Violation of the derived upper bound on the mesh Reynolds number may lead to spurious oscillations, but it may also result in less obviously recognizable errors that are visible in different ways, e.g., as incorrect predictions of regions of large variation. The limit on the mesh Reynolds number gets more severe for higher-order spatial discretization operators. This limit is also a function of the truncation order of the gPC expansion, becoming more restrictive for more accurate expansions.

In the case of spatially independent viscosity as well as spatially varying viscosity, the advection-diffusion stochastic Galerkin system can be diagonalized under some conditions. This diagonalization results in a number of uncoupled problems, and the numerical cost and performance are very similar to those of non-intrusive methods such as pseudospectral projection and stochastic collocation.

For diffusive problems, the stochastic Galerkin formulation leads to better accuracy than does stochastic collocation. For steady-state calculations, stochastic collocation is faster for advection-dominated cases and stochastic Galerkin is faster for diffusive cases. When diagonalization of the viscosity matrix $\boldsymbol{B}$ is possible, the problem should be solved in a non-intrusive way to reduce the computational cost.

SBP operators are suitable for smooth problems like the advection-diffusion equation investigated here, but many real-world flow problems contain regions of sharp gradients or discontinuities. For these problems, one may use hybrid schemes consisting of shock-capturing methods in regions of strong variation, coupled through weak interfaces with SBP schemes in smooth regions. Such methods will be investigated in the context of a two-phase problem in Chap. 9.

5.8 Supplementary Codes

Matlab scripts for the advection-diffusion equation with the stochastic Galerkin method and summation-by-parts operators can be downloaded from [http://extras.springer.com]. We encourage the reader to experiment with the scripts as a complement to the exercises.

To get started with the codes, simply run the script `advection_diffusion_main.m` in Matlab. Choose `mod='lege'` to simulate uniformly distributed viscosity using Hermite polynomials, and set `mod='herm'` to simulate lognormal viscosity with Legendre polynomials.

To maintain stability, any changes in the problem parameters should follow the derivations of this chapter. The script `SBP_operators.m` contains summation-by-parts operators of orders 2,4,6 and 8 and is a generic implementation that may be used for other problems of interest. The scripts `hermite_chaos.m` and `legendre_chaos.m` define the inner triple products $\langle \psi_i \psi_j \psi_k \rangle$ of univariate Hermite and Legendre polynomials, respectively. By replacing these scripts with the corresponding triple products for other basis functions, other classes of polynomial chaos can be used. Note that initial and boundary conditions are in general specific to the choice of basis functions.

5.9 Exercises

5.1. Consider the problem (5.1) in which μ is deterministic (i.e., $\mu = \mu(x)$ only) and V is uncertain (i.e. $v = v(\xi)$). Derive the equations for the gPC coefficients and show that the arguments illustrated in Sect. 5.1.2 apply to this case as well. What are the implications in terms of stability? Is (5.27) still valid (with the appropriate definitions)?

5.2. Solve numerically the advection-diffusion problem reported in Fig. 5.2 using a constant value of $\mu = 0.1$ and an uncertain convection speed $v = \mathcal{U}\,[0.95, 1.05]$.

5.3. Consider the case in which both the convection speed and the diffusion coefficient are uncertain but perfectly correlated $\mu = \xi$, $v = 0.9 + \xi$. Assume $\xi = \mathcal{U}[0.05, 0.15]$. Study again the stability characteristics in terms of Re_{mesh}. Solve the problem in the previous exercise and compare the results.

5.4. Extend the stochastic Galerkin framework to multiple stochastic dimensions by introducing products of single-dimensional polynomials. Do this by generalizing the computation of inner triple products $\langle \psi_i \psi_j \psi_k \rangle$ defined for a single stochastic dimension in (5.18), to hold for multiple dimensions.

Hint: you can use the provided Matlab script as a starting point, and add a loop over the single-dimension inner products to account for multiple dimensions.

5.5. The numerical results presented are all obtained by an explicit time integration scheme (fourth-order Runge-Kutta). For the steady-state problem in Sect. 5.5.4, formulate an implicit time integration method. Use the supplied Matlab codes for the advection-diffusion equation as a template and implement the implicit integration method. How do the implicit and explicit time integration methods compare in simulation time when advancing towards steady-state?

References

1. Azor R, Gillis J, Victor JD (1982) Combinatorial applications of Hermite polynomials. SIAM J Math Anal 13(5):879–890
2. Carpenter MH, Gottlieb D, Abarbanel S (1994) Time-stable boundary conditions for finite-difference schemes solving hyperbolic systems: methodology and application to high-order compact schemes. J Comput Phys 111(2):220–236. doi:http://dx.doi.org/10.1006/jcph.1994.1057
3. Coutelieris FA, Delgado JMPQ Transport processes in porous media. Springer, Berlin/New York (2012)
4. Ghanem RG (1997) Stochastic finite elements for heterogeneous media with multiple random non-Gaussian properties. ASCE J Eng Mech 125:26–40
5. Ghanem RG, Dham S (1998) Stochastic finite element analysis for multiphase flow in heterogeneous porous media. Porous Media 32:239–262
6. Golub GH, Welsch JH (1969) Calculation of Gauss quadrature rules. Math Comput 23(106):221–230. http://www.jstor.org/stable/2004418
7. Gottlieb D, Xiu D (2008) Galerkin method for wave equations with uncertain coefficients. Commun Comput Phys 3(2):505–518
8. Gustafsson B, Kreiss HO, Sundström A (1972) Stability theory of difference approximations for mixed initial boundary value problems. II. Math Comput 26(119):649–686. http://www.jstor.org/stable/2005093
9. Jameson A (1979) Acceleration of transonic potential flow calculations on arbitrary meshes by the multiple grid method. In: Proceedings of the fourth AIAA computational fluid dynamics conference, Williamsburg, July 1979. AIAA Paper 79-1458
10. Jameson A (1991) Time dependent calculations using multigrid, with applications to unsteady flows past airfoils and wings. In: Proceedings of the 10th computational fluid dynamics conference, Honolulu, 24–26 June 1991. AIAA Paper 91-1596

11. Jameson A, Baker TJ (1983) Solution of the Euler equations for complex configurations. In: Proceedings of the. 6th AIAA computational fluid dynamics conference, Danvers, July 1983. Conference proceeding series. AIAA. AIAA Paper 83-1929
12. Kreiss HO, Scherer G (1974) Finite element and finite difference methods for hyperbolic partial differential equations. In: Mathematical aspects of finite elements in partial differential equations. Academic, New York, pp 179–183
13. Le Maître OP, Knio OM, Reagan M, Najm HN, Ghanem RG (2001) A stochastic projection method for fluid flow. I: Basic formulation. J Comput Phys 173:481–511
14. Mattsson K (2012) Summation by parts operators for finite difference approximations of second-derivatives with variable coefficients. J Sci Comput 51:650–682. http://dx.doi.org/10.1007/s10915-011-9525-z
15. Mattsson K, Nordström J (2004) Summation by parts operators for finite difference approximations of second derivatives. J Comput Phys 199(2):503–540. doi:10.1016/j.jcp.2004.03.001, http://dx.doi.org/10.1016/j.jcp.2004.03.001
16. Nordström J (1989) The influence of open boundary conditions on the convergence to steady state for the Navier-Stokes equations. J Comput Phys 85(1):210–244. doi:10.1016/0021-9991(89)90205-2, http://www.sciencedirect.com/science/article/pii/0021999189902052
17. Nordström J, Eriksson S, Eliasson P (2012) Weak and strong wall boundary procedures and convergence to steady-state of the Navier-Stokes equations. J Comput Phys 231(14):4867–4884
18. Pettersson P, Doostan A, Nordström J (2013) On stability and monotonicity requirements of finite difference approximations of stochastic conservation laws with random viscosity. Comput Methods Appl Mech Eng 258(0):34–151. doi:http://dx.doi.org/10.1016/j.cma.2013.02.009
19. Pettersson P, Iaccarino G, Nordström J (2009) Numerical analysis of the Burgers' equation in the presence of uncertainty. J Comput Phys 228:8394–8412. doi:10.1016/j.jcp.2009.08.012, http://dl.acm.org/citation.cfm?id=1621150.1621394
20. Roache PJ (1988) Verification of codes and calculations. AIAA J 36(5):696–702
21. Shunn L, Ham FE, Moin P (2012) Verification of variable-density flow solvers using manufactured solutions. J Comput Phys 231(9):3801–3827
22. Sonday BE, Berry RD, Najm HN, Debusschere BJ (2011) Eigenvalues of the Jacobian of a Galerkin-projected uncertain ODE system. SIAM J Sci Comput 33:1212–1233. doi:http://dx.doi.org/10.1137/100785922, http://dx.doi.org/10.1137/100785922
23. Spijker M (1996) Stiffness in numerical initial-value problems. J Comput Appl Math 72:393–406
24. Svärd M, Nordström J (2006) On the order of accuracy for difference approximations of initial-boundary value problems. J Comput Phys 218(1):333–352. doi:10.1016/j.jcp.2006.02.014, http://dx.doi.org/10.1016/j.jcp.2006.02.014
25. Ullmann E (2008) Solution strategies for stochastic finite element discretizations. Ph.D. thesis, Technische Universität Bergakademie Freiberg, Germany
26. Wan X, Karniadakis GE (2006) Long-term behavior of polynomial chaos in stochastic flow simulations. Comput Methods Appl Math Eng 195:5582–5596
27. Wan X, Xiu D, Karniadakis GE (2005) Stochastic solutions for the two-dimensional advection-diffusion equation. SIAM J Sci Comput 26:578–590
28. Xiu D, Karniadakis GE (2002) Modeling uncertainty in steady state diffusion problems via generalized polynomial chaos. Comput Methods Appl Math Eng 191:4927–4948
29. Xiu D, Karniadakis GE (2003) Modeling uncertainty in flow simulations via generalized polynomial chaos. J Comput Phys 187(1):137–167. doi:10.1016/S0021-9991(03)00092-5, http://dx.doi.org/10.1016/S0021-9991(03)00092-5
30. Xiu D, Shen J (2009) Efficient stochastic Galerkin methods for random diffusion equations. J Comput Phys 228(2):266–281. doi:10.1016/j.jcp.2008.09.008, http://dx.doi.org/10.1016/j.jcp.2008.09.008

Chapter 6
Nonlinear Transport Under Uncertainty

Burgers' equation is a non-linear model problem from which many results can be extended to other hyperbolic systems, e.g., the Euler equations. In this chapter, a detailed uncertainty quantification analysis is performed for the Burgers' equation; we employ a spectral representation of the solution in the form of polynomial chaos expansion. The PDE is stochastic as a result of the uncertainty in the initial and boundary values. Stochastic Galerkin projection results in a coupled, deterministic system of nonlinear hyperbolic equations from which statistics of the solution can be determined.

Previous investigations on the effect of uncertainty on Burgers' equation focused on the location of the transition layer of a shock discontinuity arising in simulations of the Burgers' equation with nonzero viscosity. Small, one-sided perturbations imply large variation in the location of the transition layer, so-called *supersensitivity* [15], which has been shown to be a problem in deterministic as well as stochastic simulations. The results from the polynomial chaos approach were accurate and the method was faster than the Monte Carlo method [14, 15]. Burgers' equation with a stochastic forcing term has also been investigated and compared to standard Monte Carlo methods [6].

In this chapter, based on [12], we perform a fundamental analysis of the Burgers' equation and develop a numerical framework to study the effect of uncertainty in the boundary conditions. With the assumption that the uncertainty of the boundary data has a Gaussian distribution we allow the occurrence of unbounded solutions. Assuming that the boundary data resemble the Gaussian distribution but are bounded to a sufficiently large range does not alter the numerical results. Convergence is proven by a suitable choice of functional space.

In order to ensure stability of the discretized system of equations, SBP operators and weak imposition of boundary conditions [2, 10, 11] are used to obtain energy estimates, as demonstrated in Chap. 5. The system is expressed in a split form that combines the conservative and non-conservative formulation [9]. A particular set of

M.P. Pettersson et al., *Polynomial Chaos Methods for Hyperbolic Partial Differential Equations*, Mathematical Engineering,
DOI 10.1007/978-3-319-10714-1_6

artificial dissipation operators [8] and the simultaneous approximation term (SAT) technique [1] for boundary treatment are used to enhance stability close to the shock. The discretization method is based on a fourth-order central difference operator in space and a fourth-order Runge-Kutta method in time. The SBP operators ensure stable solutions, but the allowed time-step decreases with increasing gPC expansion as a result of the eigenvalues growing with the order of the polynomial order (i.e., the size of the system).

An analytical solution is derived for a discontinuous and uncertain initial condition: the expectation and variance of the solution are shown to be smooth functions, whereas the coefficients of truncated polynomial chaos expansions are discontinuous. Analysis of the characteristics of the truncated system also shows that the boundary values are time-dependent and suggests a way of imposing accurate boundary conditions.

In this chapter we also investigate to what extent low-order approximations can be used when appropriate high-order boundary data (i.e., data with known high-order moments) are missing. Due to the lack of boundary data as well as to the computational cost of higher-order polynomial chaos simulations, low-order approximations with appropriate utilization of available data are a viable option. Because of the hyperbolic nature of the problem, information is traveling with finite but unknown speed through the domain and will eventually affect the boundary solution values.

By the convergence properties of the polynomial chaos series expansion, higher-order boundary terms are expected to decrease rapidly. On the other hand, although small, these coefficients have a relatively large impact on the system eigenvalues and might thus be crucial for accurate boundary treatment. In addition, there are discontinuities in the stochastic dimension (we assume only one stochastic dimension), which deteriorates the convergence. The overall effect of the higher-order boundary coefficients is not clear, which provides the impetus for the investigation of this chapter.

6.1 Polynomial Chaos Expansion of Burgers' Equation

Consider the inviscid Burgers' equation in non-conservative form

$$\frac{\partial u}{\partial t} + u\frac{\partial u}{\partial x} = 0, \quad 0 \leq x \leq 1. \tag{6.1}$$

The solution $u(x,t,\xi)$ is represented as a polynomial chaos series in the set of Hermite polynomials $\{\psi_i\}_{i=0}^{\infty}$ of a standard Gaussian random variable $\xi \sim \mathcal{N}(0,1)$. The gPC series $u(x,t,\xi) = \sum_{i=0}^{\infty} u_i \psi_i(\xi)$ is inserted into (6.1), which yields

$$\sum_{i=0}^{\infty} \frac{\partial u_i}{\partial t} \psi_i(\xi) + \left(\sum_{j=0}^{\infty} u_j \psi_j(\xi) \right) \left(\sum_{i=0}^{\infty} \frac{\partial u_i}{\partial x} \psi_i(\xi) \right) = 0. \tag{6.2}$$

A stochastic Galerkin projection is performed by multiplying (6.2) by $\psi_k(\xi)$ for non-negative integers k and integrating over the probability domain Ω with respect to the Gaussian measure, i.e., with the weight function $\tilde{p}(\xi) = \exp(-\xi^2/2)/\sqrt{2\pi}$. The orthogonality of the basis polynomials then yields a system of deterministic equations. By truncating the number of polynomial chaos coefficients to a finite order M, the solution is projected onto a finite dimensional space. The result is a symmetric system of deterministic equations,

$$\frac{\partial u_k}{\partial t} + \sum_{i=0}^{M} \sum_{j=0}^{M} u_i \frac{\partial u_j}{\partial x} \langle \psi_i \psi_j \psi_k \rangle = 0 \qquad \text{for } k = 0, 1, \ldots, M. \tag{6.3}$$

For simplicity of notation, Eq. (6.3) can be written in matrix form as

$$\boldsymbol{u}_t^M + \boldsymbol{A}(\boldsymbol{u}^M)\boldsymbol{u}_x^M = \boldsymbol{0} \qquad \text{or} \qquad \boldsymbol{u}_t^M + \frac{1}{2}\frac{\partial}{\partial x}(\boldsymbol{A}(\boldsymbol{u}^M)\boldsymbol{u}^M) = \boldsymbol{0}, \tag{6.4}$$

where the matrix $\boldsymbol{A}(\boldsymbol{u}^M)$ is defined by $[\boldsymbol{A}(\boldsymbol{u}^M)]_{jk} = \sum_{i=0}^{M} \langle \psi_i \psi_j \psi_k \rangle u_i$.

6.1.1 Entropy and Energy Estimates for the $M = 2$ Case

As an illustration, the 3×3 system given by (6.4) and truncation of the expansion to $M = 2$ with a normalized Hermite polynomial basis is

$$\begin{pmatrix} u_0 \\ u_1 \\ u_2 \end{pmatrix}_t + \begin{pmatrix} u_0 & u_1 & u_2 \\ u_1 & u_0 + \sqrt{2}u_2 & \sqrt{2}u_1 \\ u_2 & \sqrt{2}u_1 & u_0 + 2\sqrt{2}u_2 \end{pmatrix} \begin{pmatrix} u_0 \\ u_1 \\ u_2 \end{pmatrix}_x = 0.$$

Note that the matrix $\boldsymbol{A}(\boldsymbol{u}^M)$ is symmetric. Let $\boldsymbol{f} = \frac{1}{2}\boldsymbol{A}(\boldsymbol{u})\boldsymbol{u}$ denote the flux function of the $M = 2$ system and introduce the *entropy flux* $F = \boldsymbol{u}^T \boldsymbol{f} - G$, where

$$G = \frac{1}{6}u_0^3 + \frac{1}{2}u_0 u_1^2 + \frac{1}{2}u_0 u_2^2 + \frac{\sqrt{2}}{2}u_1^2 u_2 + \frac{\sqrt{2}}{3}u_2^3,$$

i.e., $\left(\frac{\partial G}{\partial \boldsymbol{u}}\right)^T = \boldsymbol{f}$. Then, introducing the convex entropy function $h = \frac{1}{2}\boldsymbol{u}^T \boldsymbol{u}$, and assuming smoothness, we obtain

$$h(\boldsymbol{u})_t + F(\boldsymbol{u})_x = 0.$$

Anticipating the more general time-stability analysis of Sect. 6.4, we now consider a semidiscretized formulation on an equidistant mesh with cell size Δx and solution $\boldsymbol{u}_j = \boldsymbol{u}(x_j) = (u_0(x_j,t)\ u_1(x_j,t)\ u_2(x_j,t))^T$. With the straight-line parameterization $\hat{\boldsymbol{u}}_j(\theta) = \boldsymbol{u}_j + \theta(\boldsymbol{u}_{j+1} - \boldsymbol{u}_j)$, $\theta \in [0,1]$, and flux $\hat{\boldsymbol{f}}_{j+1/2} = \int_0^1 \boldsymbol{f}(\hat{\boldsymbol{u}}_j(\theta))d\theta$, the semidiscrete formulation is given by

$$\frac{d\hat{\boldsymbol{u}}_j}{dt} + \frac{\hat{f}_{j+1/2} - \hat{f}_{j-1/2}}{\Delta x} = 0.$$

Multiplication by $\Delta x \hat{\boldsymbol{u}}_j^T$ and summing with respect to the grid index j, gives the semidiscrete energy estimate

$$\frac{d}{dt}\left(\frac{1}{2}\Delta x \sum_j \hat{\boldsymbol{u}}_j^T \hat{\boldsymbol{u}}_j\right) = \Delta x \sum_j \hat{\boldsymbol{u}}_j^T \frac{d\hat{\boldsymbol{u}}_j}{dt} = -\sum_j \hat{\boldsymbol{u}}_j^T \left(\hat{f}_{j+1/2} - \hat{f}_{j-1/2}\right) =$$

$$\sum_j \left(\hat{\boldsymbol{u}}_{j+1} - \hat{\boldsymbol{u}}_j\right)^T \hat{f}_{j+1/2} + B.T. = \sum_j G_{j+1} - G_j + B.T. = B.T., \qquad (6.5)$$

which is an expression involving the boundary terms (B.T.) only. For the $M = 2$ case, we define

$$\overline{u^2_{k,j+\frac{1}{2}}} = \frac{u^2_{k,j+1} + u_{k,j}u_{k,j+1} + u^2_{k,j}}{3}, \qquad (6.6)$$

$$\overline{u_{k,j+\frac{1}{2}}u_{l,j+\frac{1}{2}}} = \frac{2u_{k,j}u_{l,j} + u_{k,j}u_{l,j+1} + u_{k,j+1}u_{l,j} + 2u_{k,j+1}u_{l,j+1}}{6}, \qquad (6.7)$$

for $k, l = 0, 1, 2$. Then, the numerical flux function is given by

$$\hat{\boldsymbol{f}}_{j+\frac{1}{2}} = \begin{bmatrix} \frac{1}{2}\left(\overline{u^2_{0,j+\frac{1}{2}}} + \overline{u^2_{1,j+\frac{1}{2}}} + \overline{u^2_{2,j+\frac{1}{2}}}\right) \\ \overline{u_{0,j+\frac{1}{2}}u_{1,j+\frac{1}{2}}} + \sqrt{2}\overline{u_{1,j+\frac{1}{2}}u_{2,j+\frac{1}{2}}} \\ \overline{u_{0,j+\frac{1}{2}}u_{2,j+\frac{1}{2}}} + \frac{\sqrt{2}}{2}\overline{u^2_{1,j+\frac{1}{2}}} + \sqrt{2}\overline{u^2_{2,j+\frac{1}{2}}} \end{bmatrix}.$$

In Sect. 6.4, stability analysis is performed for the case of general order of expansion M.

6.1.2 *Diagonalization of the System Matrix $A(u^M)$*

For purposes such as analysis of well-posedness, design of dissipation operators and analysis of characteristics, the matrix $\boldsymbol{A}(\boldsymbol{u}^M)$ is diagonalized. This is possible for any $\boldsymbol{u}^M \in \mathbb{R}^{M+1}$ since $\boldsymbol{A}(\boldsymbol{u}^M)$ is always symmetric and thus has real-valued eigenvalues and eigenvectors.

For any given $\boldsymbol{u} \in \mathbb{R}^{M+1}$, let $\boldsymbol{\Lambda}$ denote a diagonal matrix with the eigenvalues λ_i of $\boldsymbol{A}(u)$ on the main diagonal and let $\boldsymbol{V}$ be the matrix where the columns are the linearly independent eigenvectors. Then $\boldsymbol{A}(\boldsymbol{u}^M) = \boldsymbol{V}\boldsymbol{\Lambda}\boldsymbol{V}^T$. Using the eigenvalue decomposition and momentarily assuming a linearized Burgers' equation (i.e., the speed of propagation of the waves is assumed to be constant), we obtain the diagonalized system

$$\boldsymbol{w}_t^M + \boldsymbol{\Lambda}\boldsymbol{w}_x^M = 0,$$

where $\boldsymbol{w}^M = \boldsymbol{V}^T\boldsymbol{u}^M$. Assuming nonzero eigenvalues, $\boldsymbol{\Lambda}$ can be split according to the sign of its eigenvalues as $\boldsymbol{\Lambda} = \boldsymbol{\Lambda}^+ + \boldsymbol{\Lambda}^-$. Introducing the split scheme into the system of equations gives

$$\boldsymbol{w}_t^M + \boldsymbol{\Lambda}^+\boldsymbol{w}_x^M + \boldsymbol{\Lambda}^-\boldsymbol{w}_x^M = 0. \tag{6.8}$$

This form will be used in the following sections.

6.2 A Reference Solution

In order to quantify the accuracy of the numerical methods, we need an analytical solution to our problem. Consider the stochastic Riemann problem with an initial shock of uncertain strength located at $x_0 \in [0, 1]$

$$\begin{aligned} u(x,0,\xi) &= \begin{cases} u_L = a + p(\xi) & \text{if } x < x_0 \\ u_R = -a + p(\xi) & \text{if } x > x_0 \end{cases} \\ u(0,t,\xi) &= u_L,\ u(1,t,\xi) = u_R \\ \xi &\in \mathcal{N}(0,1). \end{aligned} \tag{6.9}$$

As the most intuitive choice of polynomial chaos basis with regard to the uncertainty in the initial and boundary conditions, the set of Hermite polynomials will be used. Here we will only consider $p(\xi) = b\xi$ as a first-order stochastic polynomial and a is a constant. By the Rankine-Hugoniot condition, the shock speed is given by $s = b\xi$, so for any bounded ξ the shock location x_s is

$$x_s = x_0 + tb\xi.$$

The solution (for any bounded ξ) is given by

$$u(x,t,\xi) = \begin{cases} u_L \text{ if } x < x_0 + tb\xi \\ u_R \text{ if } x > x_0 + tb\xi. \end{cases}$$

Since the analytical solution is known, the coefficients of the complete gPC expansion ($M \to \infty$) can be calculated for any given i, x and t. We have

$$u_i(x,t) = \int_{-\infty}^{\infty} u(x,t,\xi)\psi_i(\xi)\tilde{p}(\xi)d\xi = a\delta_{i0} + b\delta_{i1} - 2a\int_{-\infty}^{\xi_s} \psi_i\tilde{p}(\xi)d\xi, \tag{6.10}$$

where we have defined $\xi_s = (x - x_0)/(bt)$ and $\tilde{p}(\xi) = \exp(-\xi^2/2)/\sqrt{2\pi}$ denotes the Gaussian probability density function. Note that the limit of integration $\xi_s(x,t)$ is not a random variable itself. Using the recursion relation for normalized Hermite polynomials

$$\psi_i(\xi) = \frac{1}{\sqrt{i}}\left(\xi\psi_{i-1}(\xi) - \psi'_{i-1}(\xi)\right),$$

(6.10) can be written

$$u_i(x,t) = b\delta_{i1} + a\sqrt{\frac{2}{i\pi}}\psi_{i-1}(\xi_s)e^{-\xi_s^2/2}, \tag{6.11}$$

for $i \geq 1$. Differentiating (6.11) with respect to x and t results in

$$\frac{\partial u_i}{\partial x} = \frac{\partial u_i}{\partial \xi_s}\frac{\partial \xi_s}{\partial x} = -2a\psi_i(\xi_s(x,t))\tilde{p}(\xi_s(x,t))\frac{1}{bt},$$

and

$$\frac{\partial u_i}{\partial t} = \frac{\partial u_i}{\partial \xi_s}\frac{\partial \xi_s}{\partial t} = 2a\psi_i(\xi_s(x,t))\tilde{p}(\xi_s(x,t))\frac{x - x_0}{bt^2},$$

from which it is clear that $u_i(x,t)$ is continuous in x and t for $x \in [0,1]$ and $t > 0$. (The same is true for u_0.) With an appropriate choice of initial function, the coefficients would also be continuous for $t = 0$. (For treatment of a similar case of smooth coefficients of a discontinuous solution, see [3]).

The effect of the introduction of a finite series in representing the solution will be studied next in terms of comparisons to the expected value and the variance of the analytical solution, given by

$$E(u) = a\left(1 - 2\int_{-\infty}^{\xi_s(x,t)} \frac{e^{-\xi^2/2}}{\sqrt{2\pi}}d\xi\right), \tag{6.12}$$

and

$$Var(u) = b^2 + 4ab\frac{e^{-\xi_s^2/2}}{\sqrt{2\pi}} + 4a^2\int_{-\infty}^{\xi_s}\frac{e^{-\xi^2/2}}{\sqrt{2\pi}}d\xi - 4a^2\left(\int_{-\infty}^{\xi_s}\frac{e^{-\xi^2/2}}{\sqrt{2\pi}}d\xi\right)^2. \tag{6.13}$$

These expressions can be generalized for different boundary conditions and polynomial bases.

6.2.1 Regularity Determined by the gPC Expansion Order

The solution of (6.1) is obtained through the evaluation of the gPC series with the coefficients given by (6.10). Figure 6.1 shows the solution obtained by retaining only (a) the zeroth- and (b) first-order gPC expansion terms. As a contrast to these low-order approximations, the true solution is discontinuous, as shown in (c). However, due to the nonlinearities, the finite order M solution of the truncated stochastic Galerkin system (6.4) is not equal to the order M solution of the original problem defined by the coefficients (6.10). The exact solutions of the zeroth-, first- and second-order stochastic Galerkin systems are shown in Fig. 6.2. Unlike the smooth coefficients of the original problem, the solutions (and coefficients) of the truncated systems are discontinuous.

The dependence of smoothness on the order of gPC expansion is illustrated in Fig. 6.3, where the expectation is shown as a function of space for different orders of gPC expansion and fixed time $t = 0.5$. For $M = 0, 1, 2$, there are, respectively, 1,2 or 3 solution discontinuities of the expectations. In Fig. 6.3d, the $M = 3$ expectation appears to exhibit an expansion wave.

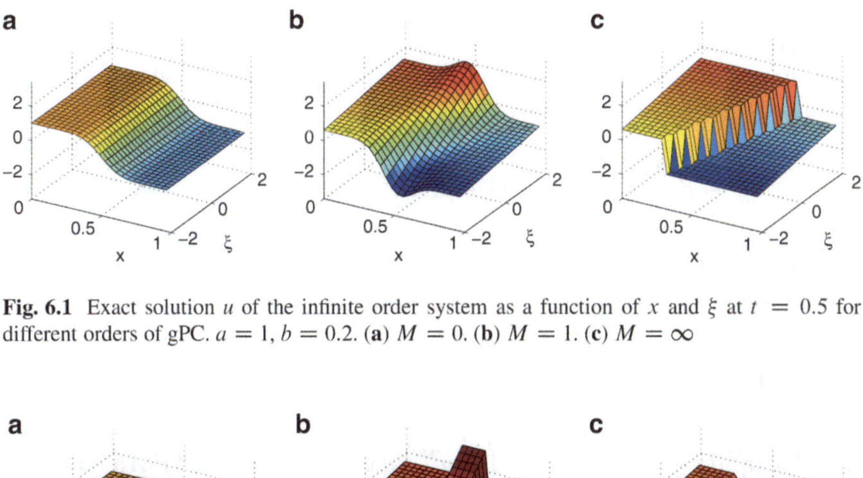

Fig. 6.1 Exact solution u of the infinite order system as a function of x and ξ at $t = 0.5$ for different orders of gPC. $a = 1, b = 0.2$. (**a**) $M = 0$. (**b**) $M = 1$. (**c**) $M = \infty$

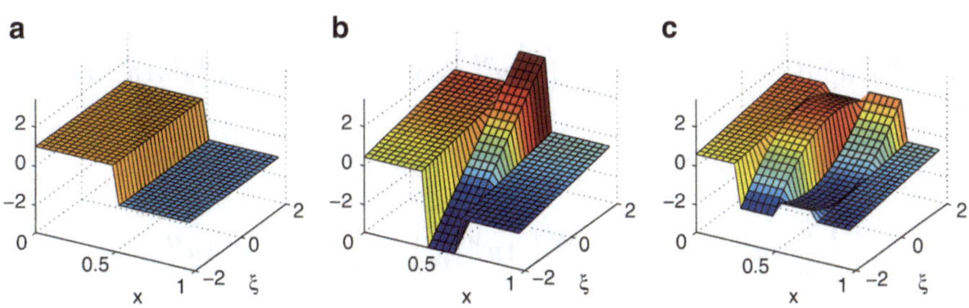

Fig. 6.2 Exact solution u of the truncated system as a function of x and ξ at $t = 0.5$ for different orders of gPC. $a = 1, b = 0.2$. (**a**) $M = 0$. (**b**) $M = 1$. (**c**) $M = 2$

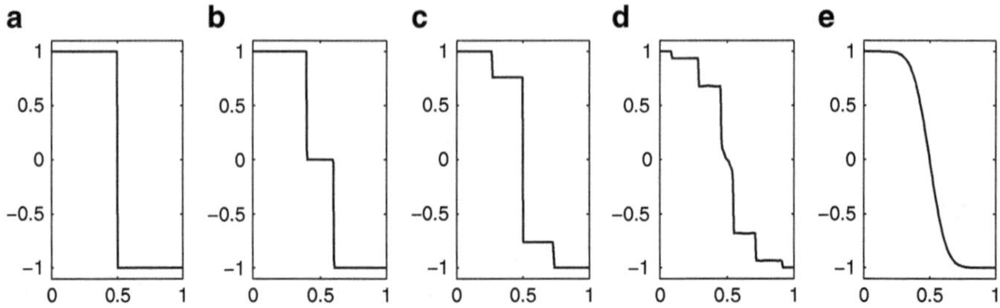

Fig. 6.3 Expectation u_0 as a function of x at $t = 0.5$ for different orders of gPC. $a = 1, b = 0.2$. (**a**) $M = 0$. (**b**) $M = 1$. (**c**) $M = 2$. (**d**) $M = 3$. (**e**) $M = \infty$

6.3 Well-Posedness

The solution of (6.4) requires initial and boundary data. The data depend on the expected conditions and the distribution of the uncertainty introduced; the stochastic Galerkin procedure is again used to determine the polynomial chaos coefficients for the initial and boundary values. In this section we will show that the truncated system resulting from a truncated gPC expansion is well-posed if the correct boundary conditions are given.

In the rest of this section, we assume $\boldsymbol{u}^M$ to be sufficiently smooth. Consider the continuous problem in split form [13]

$$\boldsymbol{u}_t^M + \beta \frac{\partial}{\partial x}\left(\frac{\boldsymbol{A}}{2}\boldsymbol{u}^M\right) + (1-\beta)\boldsymbol{A}\boldsymbol{u}_x^M = 0,\ 0 \le x \le 1. \tag{6.14}$$

Proposition 6.1. *The split form of Burgers' equation (6.14) with the weight $\beta = 2/3$ is strongly well-posed in the sense of Definition 1.2*

Proof. Multiplication of (6.14) by $(\boldsymbol{u}^M)^T$ and integration over the spatial domain $\Omega_{\text{phys}} = [0, 1]$ yields

$$\int_0^1 (\boldsymbol{u}^M)^T \boldsymbol{u}_t^M dx + \beta \int_0^1 (\boldsymbol{u}^M)^T \frac{\partial}{\partial x}\left(\frac{\boldsymbol{A}}{2}\boldsymbol{u}^M\right) dx + (1-\beta)\int_0^1 (\boldsymbol{u}^M)^T \boldsymbol{A}\boldsymbol{u}_x^M dx = 0.$$

Integration by parts gives

$$\begin{aligned}\frac{1}{2}\frac{\partial}{\partial t}\left\|\boldsymbol{u}^M\right\|^2 = &-\frac{\beta}{2}[(\boldsymbol{u}^M)^T \boldsymbol{A}\boldsymbol{u}^M]_{x=0}^{x=1} + \frac{\beta}{2}\int_0^1 (\boldsymbol{u}_x^M)^T \boldsymbol{A}\boldsymbol{u}^M dx \\ &-(1-\beta)\int_0^1 (\boldsymbol{u}^M)^T \boldsymbol{A}\boldsymbol{u}_x^M dx.\end{aligned} \tag{6.15}$$

We choose β such that

$$\frac{\beta}{2} - (1-\beta) = 0 \Leftrightarrow \beta = \frac{2}{3},$$

which is inserted into (6.15), yielding

$$\begin{aligned}\frac{\partial}{\partial t}\left\|\boldsymbol{u}^M\right\|^2 &= -\frac{2}{3}[(\boldsymbol{u}^M)^T \boldsymbol{A}\boldsymbol{u}^M]_{x=0}^{x=1} \\ &= \frac{2}{3}\left((\boldsymbol{w}_0^M)^T(\boldsymbol{\Lambda}_0^+ + \boldsymbol{\Lambda}_0^-)\boldsymbol{w}_0^M - (\boldsymbol{w}_1^M)^T(\boldsymbol{\Lambda}_1^+ + \boldsymbol{\Lambda}_1^-)\boldsymbol{w}_1^M\right), \qquad (6.16)\end{aligned}$$

where $\boldsymbol{A}(\boldsymbol{u}^M)$ has been diagonalized at the boundaries according to Sect. 6.1.2. Boundary conditions are imposed on the resulting incoming characteristic variables which correspond to $\boldsymbol{\Lambda}^+$ for $x = 0$ and $\boldsymbol{\Lambda}^-$ for $x = 1$. On the left boundary, the conditions are set such that,

$$(\boldsymbol{w}_0^M)_i = (\boldsymbol{V}^T\boldsymbol{u}^M(x=0))_i = (\boldsymbol{g}_L^M)_i \text{ if } \lambda_i > 0$$

and on the right boundary,

$$(\boldsymbol{w}_1^M)_i = (\boldsymbol{V}^T u(x=1))_i = (\boldsymbol{g}_R^M)_i \text{ if } \lambda_i < 0.$$

The boundary norm is defined as

$$\begin{aligned}\left\|\boldsymbol{w}^M\right\|_{\Gamma_{\text{phys}}}^2 &= (\boldsymbol{w}^M)^T\boldsymbol{\Lambda}^+\boldsymbol{w}^M - (\boldsymbol{w}^M)^T\boldsymbol{\Lambda}^-\boldsymbol{w}^M = (\boldsymbol{w}^M)^T(\boldsymbol{\Lambda}^+ + |\boldsymbol{\Lambda}^-|)\boldsymbol{w}^M \\ &= (\boldsymbol{w}^M)^T\left|\boldsymbol{\Lambda}\right|\boldsymbol{w}^M \text{ for } x = 0, 1. \qquad (6.17)\end{aligned}$$

Inserting the boundary conditions and integrating (6.16) over time gives

$$\begin{aligned}&\left\|\boldsymbol{u}^M\right\|_{\Omega_{\text{phys}}}^2 + \frac{2}{3}\int_0^t \left\|\boldsymbol{w}_0^M\right\|_{\Gamma_{\text{phys}}}^2 + \left\|\boldsymbol{w}_1^M\right\|_{\Gamma_{\text{phys}}}^2 d\tau \le \\ &\quad\le \left\|\boldsymbol{u}^M(t=0)\right\|_{\Omega_{\text{phys}}}^2 + \frac{4}{3}\int_0^t \left\|\boldsymbol{g}_L^M\right\|_{\Gamma_{\text{phys}}}^2 + \left\|\boldsymbol{g}_R^M\right\|_{\Gamma_{\text{phys}}}^2 d\tau. \qquad (6.18)\end{aligned}$$

Since $\left\|\boldsymbol{w}^M\right\| \le \left\|\boldsymbol{V}^T\right\| \left\|\boldsymbol{u}^M\right\| \le C\left\|\boldsymbol{u}^M\right\|$ for some $C < \infty$, the estimate (6.18) is in the form of Eq. (1.3) and the problem is strongly well-posed according to Definition 1.2.

Remark 6.1. The assumption that $\boldsymbol{u}^M$ is smooth is actually true for an infinite number of terms of the polynomial chaos expansion and $t > 0$.

6.3.1 *The Importance of Boundary Conditions*

We have seen that the imposition of suitable boundary conditions is crucial for analysis of well-posedness. In fact, we even formulated the problem as one of finding the correct boundary conditions for a given problem, rather than starting from given boundary conditions and then trying to prove well-posedness. Whether we start from a linear or a nonlinear problem, a scalar equation or a system, the procedure is the same: identify ingoing and outgoing characteristics, then impose boundary conditions on ingoing waves such that growth terms are controlled. The situation is analogous for the discrete system. In order to maintain stability – the discrete analogue of well-posedness, a correct number of boundary conditions must be imposed.

When nonlinear waves are crossing the boundaries of the spatial domain of interest, the Jacobian of the flux function may change and lead to an increased or decreased number of ingoing waves. The result is that the number of boundary conditions changes. It is therefore important to keep track of the evolution of the system along the boundaries. In general, one has to evaluate the Jacobian numerically to keep track of the in- and outgoing waves.

The Jacobian at a boundary point of interest may be positive definite for certain orders of gPC expansions, but not for others. This phenomenon was investigated in detail in [5] and also pertains to linear problems since it determines the number of boundary conditions required for well-posedness.

6.4 Energy Estimates for Stability

In order to ensure stability of the discretized system of equations, SBP operators and weak imposition of boundary conditions [1, 2, 10, 11] are used to obtain energy estimates. A particular set of artificial dissipation operators [8] are used to enhance stability close to the shock. Burgers' equation has been discretized with a fourth-order central difference operator in space and a fourth-order Runge-Kutta method in time. Using the provided scripts, the reader can set the spatial order of accuracy. For stability, artificial dissipation is added based on the local system eigenvalues. The order of accuracy is not affected by the addition of artificial dissipation.

The case of interest corresponds to sufficiently low artificial dissipation such that the dominating error is due to truncation of the polynomial chaos expansion. General difficulties related to solving hyperbolic problems and nonlinear conservation laws with spectral methods, to which the gPC methods belong, are discussed in [4].

In the following we use the notation and definitions of the SBP operators introduced in Chap. 4. To obtain stability, we will use the *penalty technique* [8] to impose boundary conditions for the discrete problem [12]. Assume an equidistant spatial mesh with m mesh points $x_1 = 0, x_2 = \Delta x, \ldots, x_m = 1$, where $\Delta x = 1/(m-1)$. Let $\boldsymbol{E}_1 = (e_{ij})$ where $e_{11} = 1, e_{ij} = 0, \forall i, j \neq 1$ and $\boldsymbol{E}_m = (e_{ij})$ where $e_{mm} = 1, e_{ij} = 0, i, j \neq m$. Define the block diagonal matrix $\boldsymbol{A}_g$ where the diagonal blocks are the symmetric matrices $\boldsymbol{A}(\boldsymbol{u}^M(x_i))$, $i = 1, \ldots, m$. With penalty matrices $\boldsymbol{\Sigma}_L$ and $\boldsymbol{\Sigma}_R$ corresponding to the left and right boundaries, respectively, the discretized system can be expressed as

$$\vec{\boldsymbol{u}}_t + \boldsymbol{A}_g(\boldsymbol{P}^{-1}\boldsymbol{Q} \otimes \boldsymbol{I})\vec{\boldsymbol{u}} = (\boldsymbol{P}^{-1} \otimes \boldsymbol{I})(\boldsymbol{E}_1 \otimes \boldsymbol{\Sigma}_L)(\vec{\boldsymbol{u}} - \vec{\boldsymbol{g}}_L) + (\boldsymbol{P}^{-1} \otimes \boldsymbol{I})(\boldsymbol{E}_m \otimes \boldsymbol{\Sigma}_R)(\vec{\boldsymbol{u}} - \vec{\boldsymbol{g}}_R). \quad (6.19)$$

Similarly, the conservative system in (6.4) can be discretized as

$$\vec{\boldsymbol{u}}_t + \frac{1}{2}(\boldsymbol{P}^{-1}\boldsymbol{Q} \otimes \boldsymbol{I})\boldsymbol{A}_g\vec{\boldsymbol{u}} = (\boldsymbol{P}^{-1} \otimes \boldsymbol{I})(\boldsymbol{E}_1 \otimes \boldsymbol{\Sigma}_L)(\vec{\boldsymbol{u}} - \vec{\boldsymbol{g}}_L) + (\boldsymbol{P}^{-1} \otimes \boldsymbol{I})(\boldsymbol{E}_m \otimes \boldsymbol{\Sigma}_R)(\vec{\boldsymbol{u}} - \vec{\boldsymbol{g}}_R). \quad (6.20)$$

Neither of the formulations (6.19) nor (6.20) will lead to an energy estimate. However, the non-conservative and conservative forms can be combined to get an energy estimate by using the summation by parts property. The split form is given by

$$\vec{\boldsymbol{u}}_t + \beta\frac{1}{2}(\boldsymbol{P}^{-1}\boldsymbol{Q} \otimes \boldsymbol{I})\boldsymbol{A}_g\vec{\boldsymbol{u}} + (1-\beta)\boldsymbol{A}_g(\boldsymbol{P}^{-1}\boldsymbol{Q} \otimes \boldsymbol{I})\vec{\boldsymbol{u}} = (\boldsymbol{P}^{-1} \otimes \boldsymbol{I})\left[(\boldsymbol{E}_1 \otimes \boldsymbol{\Sigma}_L)(\vec{\boldsymbol{u}} - \vec{\boldsymbol{g}}_L) + (\boldsymbol{E}_m \otimes \boldsymbol{\Sigma}_R)(\vec{\boldsymbol{u}} - \vec{\boldsymbol{g}}_R)\right]. \quad (6.21)$$

Proposition 6.2. *The linear combination of conservative and non-conservative semidiscretization (6.21) is stable with the weight $\beta = 2/3$.*

Proof. Multiplication of (6.21) from the left by $\vec{\boldsymbol{u}}^T(\boldsymbol{P} \otimes \boldsymbol{I})$ and then addition of the transpose of the resulting equation yields

$$\frac{\partial}{\partial t}\|\vec{\boldsymbol{u}}\|^2_{(\boldsymbol{P}\otimes\boldsymbol{I})} + \frac{\beta}{2}\vec{\boldsymbol{u}}^T\left((\boldsymbol{Q} \otimes \boldsymbol{I})\boldsymbol{A}_g + \boldsymbol{A}_g(\boldsymbol{Q}^T \otimes \boldsymbol{I})\right)\vec{\boldsymbol{u}} + (1-\beta)\vec{\boldsymbol{u}}^T\left(\boldsymbol{A}_g(\boldsymbol{Q} \otimes \boldsymbol{I}) + (\boldsymbol{Q}^T \otimes \boldsymbol{I})\boldsymbol{A}_g\right)\vec{\boldsymbol{u}} = 2\vec{\boldsymbol{u}}^T(\boldsymbol{E}_1 \otimes \boldsymbol{\Sigma}_L)(\vec{\boldsymbol{u}} - \vec{\boldsymbol{g}}_L) + 2\vec{\boldsymbol{u}}^T(\boldsymbol{E}_m \otimes \boldsymbol{\Sigma}_R)(\vec{\boldsymbol{u}} - \vec{\boldsymbol{g}}_R). \quad (6.22)$$

With the choice $\beta = 2/3$, the energy methods yields

$$\frac{\partial}{\partial t}\|\vec{\boldsymbol{u}}\|^2_{(\boldsymbol{P}\otimes\boldsymbol{I})} = \frac{2}{3}\left(\vec{\boldsymbol{u}}^T_{x=0}\boldsymbol{A}\vec{\boldsymbol{u}}_{x=0} - \vec{\boldsymbol{u}}^T_{x=1}\boldsymbol{A}\vec{\boldsymbol{u}}_{x=1}\right) + 2\vec{\boldsymbol{u}}^T_{x=0}\boldsymbol{\Sigma}_L(\vec{\boldsymbol{u}}_{x=0} - \vec{\boldsymbol{g}}_L) + 2\vec{\boldsymbol{u}}^T_{x=1}\boldsymbol{\Sigma}_R(\vec{\boldsymbol{u}}_{x=1} - \vec{\boldsymbol{g}}_R). \quad (6.23)$$

Restructuring (6.23) yields

$$\frac{\partial}{\partial t}\|\vec{\boldsymbol{u}}\|^2_{(\boldsymbol{P}\otimes\boldsymbol{I})} = \vec{\boldsymbol{u}}^T_{x=0}\left(\frac{2}{3}\boldsymbol{A} + 2\boldsymbol{\Sigma}_L\right)\vec{\boldsymbol{u}}_{x=0} - 2\vec{\boldsymbol{u}}^T_{x=0}\boldsymbol{\Sigma}_L\vec{\boldsymbol{g}}_L - \vec{\boldsymbol{u}}^T_{x=1}\left(\frac{2}{3}\boldsymbol{A} - 2\boldsymbol{\Sigma}_R\right)\vec{\boldsymbol{u}}_{x=1} - 2\vec{\boldsymbol{u}}^T_{x=1}\boldsymbol{\Sigma}_R\vec{\boldsymbol{g}}_R. \quad (6.24)$$

Stability is achieved by a proper choice of the penalty matrices $\boldsymbol{\Sigma}_L$ and $\boldsymbol{\Sigma}_R$. For that purpose, $\boldsymbol{A}$ is split according to the sign of its eigenvalues as

$$\boldsymbol{A} = \boldsymbol{A}^+ + \boldsymbol{A}^- \text{ where } \boldsymbol{A}^+ = \boldsymbol{V}^T \boldsymbol{\Lambda}^+ \boldsymbol{V} \text{ and } \boldsymbol{A}^- = \boldsymbol{V}^T \boldsymbol{\Lambda}^- \boldsymbol{V}. \tag{6.25}$$

Choose $\boldsymbol{\Sigma}_L$ and $\boldsymbol{\Sigma}_R$ such that $\frac{2}{3}\boldsymbol{A}^+ + 2\boldsymbol{\Sigma}_L = -\frac{2}{3}\boldsymbol{A}^+ \Leftrightarrow \boldsymbol{\Sigma}_L = -\frac{2}{3}\boldsymbol{A}^+$ and $\frac{2}{3}\boldsymbol{A}^- - 2\boldsymbol{\Sigma}_R = \frac{2}{3}\boldsymbol{A}^- \Leftrightarrow \boldsymbol{\Sigma}_R = \frac{2}{3}\boldsymbol{A}^-$. We now get the energy estimate

$$\begin{aligned}
&\frac{\partial}{\partial t}\|\vec{\boldsymbol{u}}\|^2_{(\boldsymbol{P}\otimes\boldsymbol{I})} \\
&= -\frac{2}{3}(\vec{\boldsymbol{u}}_{x=0} - \vec{\boldsymbol{g}}_L)^T \boldsymbol{A}^+ (\vec{\boldsymbol{u}}_{x=0} - \vec{\boldsymbol{g}}_L) + \frac{2}{3}\left[\vec{\boldsymbol{u}}^T_{x=0}\boldsymbol{A}^-\vec{\boldsymbol{u}}_{x=0} + \vec{\boldsymbol{g}}^T_L \boldsymbol{A}^+ \vec{\boldsymbol{g}}_L\right] \\
&\quad -\frac{2}{3}\left[\vec{\boldsymbol{u}}^T_{(x=1)}\boldsymbol{A}^+\vec{\boldsymbol{u}}_{(x=1)} + \vec{\boldsymbol{g}}^T_R \boldsymbol{A}^- \vec{\boldsymbol{g}}_R\right] + \frac{2}{3}(\vec{\boldsymbol{u}}_{(x=1)} - \vec{\boldsymbol{g}}_R)^T \boldsymbol{A}^- (\vec{\boldsymbol{u}}_{(x=1)} - \vec{\boldsymbol{g}}_R),
\end{aligned} \tag{6.26}$$

which shows that the system is strongly stable according to Definition 1.4.

Remark 6.2. In the numerical calculations we use (6.20) for correct shock speed (see [7]).

In the analysis of well-posedness and stability above we have assumed that we have perfect knowledge of boundary data, but in practice this is rarely true. Limited knowledge forces us to rely on estimates to assign boundary data. We will investigate the effect of that problem in Sect. 7.1.

6.4.1 Artificial Dissipation for Enhanced Stability

The complete difference approximation (6.20) augmented with artificial dissipation of the form described in Sect. 4.2.5 is given by

$$\begin{aligned}
(\boldsymbol{P}\otimes\boldsymbol{I})\vec{\boldsymbol{u}}_t + \frac{1}{2}(\boldsymbol{Q}\otimes\boldsymbol{I})\boldsymbol{A}_g\vec{\boldsymbol{u}} - (\boldsymbol{E}_1\otimes\boldsymbol{\Sigma}_L)(\vec{\boldsymbol{u}} - \vec{\boldsymbol{g}}_L) - (\boldsymbol{E}_m\otimes\boldsymbol{\Sigma}_R)(\vec{\boldsymbol{u}} - \vec{\boldsymbol{g}}_R) = \\
= -\Delta x \sum_k (\tilde{\boldsymbol{D}}^T_k \otimes \boldsymbol{B})\boldsymbol{B}_{w,k}(\tilde{\boldsymbol{D}}_k \otimes \boldsymbol{I})\vec{\boldsymbol{u}},
\end{aligned} \tag{6.27}$$

where $\boldsymbol{B}_{w,k}$ is a possibly non-constant weight matrix to be determined and $k = 1, 2$ for the fourth-order accurate SBP operator.

Determining $\boldsymbol{B}_{w,k}$ in (6.27) requires estimates of the eigenvalues λ_j of $\boldsymbol{A}$ for $j = 0, \ldots, M$; the largest eigenvalue is typically sufficient. For the system of equations generated by polynomial chaos expansion of Burgers' equation, $\max|\lambda|$ is not always known. Since the only nonzero polynomial coefficients on the boundaries are u_0 and u_1 and since the polynomial chaos expansion converges in the $L_2(\Omega, \mathscr{P})$ sense, a reasonable approximation of the maximum eigenvalue of $\boldsymbol{A}$ with standard (i.e., non-normalized) Hermite polynomials is

$$|\lambda|_{max} \approx |u_0| + M\,|u_1|\,, \tag{6.28}$$

where M is the order of PC expansion. This estimate is justified by the eigenvalue analysis performed in the next section, as well as by computational results. For the dissipation operators to be combined with fourth-order SBP operators in the simulations, we use

$$\boldsymbol{B}_{w,1} = \mathrm{diag}\left(\frac{(|u_0| + M\,|u_1|)}{6\Delta x}\right), \quad \boldsymbol{B}_{w,2} = \mathrm{diag}\left(\frac{(|u_0| + M\,|u_1|)}{24\Delta x}\right). \tag{6.29}$$

The second-order dissipation operator is only applied close to discontinuities. Using, say, sixth-order SBP operators, we would also need to define $\boldsymbol{B}_{w,3}$, and analogously for higher-order operators.

6.5 Time Integration

The increase in simulation cost associated with higher-order systems is attributable to a number of factors. The size of the system depends on both the number of terms in the truncated PC expansion and the spatial mesh size.

For the Kronecker product $\boldsymbol{A} \otimes \boldsymbol{B}$ the relation

$$\lambda^{i,j}_{\boldsymbol{A}\otimes\boldsymbol{B}} = \lambda^{i}_{\boldsymbol{A}}\lambda^{j}_{\boldsymbol{B}} \tag{6.30}$$

holds, where the indices i, j denote all the eigenvalues of $\boldsymbol{A}$ and $\boldsymbol{B}$, respectively. This enables a separate analysis of the eigenvalues corresponding to the PC expansion and the eigenvalues of the total spatial difference operator $\boldsymbol{D}$. Assuming constant coefficients, the maximum system eigenvalue is limited by

$$\lambda_{\max} \leq (\max \lambda_{\boldsymbol{D}})(\max \lambda_{\boldsymbol{A}}). \tag{6.31}$$

The estimate (6.31) in combination with (6.28) will be used to obtain estimates of the time-step constraint.

6.6 Eigenvalue Approximation

Analytic eigenvalues for the matrix $\boldsymbol{A}$ can be obtained only for a small number of PC coefficients and therefore approximations are needed. Even though the eigenvalues of interest in this chapter can be calculated exactly for every particular case, a general estimate is of interest. The approximation of the largest eigenvalue of the system matrix $\boldsymbol{A}$ is calculated from the solution values on the boundaries, which are the only values known a priori. For smooth solutions with boundary conditions where the PC coefficients u_i are equal to 0 for $i > 1$, the higher-order coefficients tend to remain small compared to lower-order coefficients (strong probabilistic convergence). For solutions where a shock is developing, higher-order polynomial chaos coefficients might grow and the approximation of the largest eigenvalue based on boundary values is likely to be a less accurate estimate.

To obtain estimates of the eigenvalues, the system of equations can be written

$$\boldsymbol{u}_t^M + \left(\sum_{i=0}^{M} \boldsymbol{A}_i u_i \right) \boldsymbol{u}_x^M = 0, \tag{6.32}$$

where $\boldsymbol{A}(\boldsymbol{u}^M) = \sum_{i=0}^{M} \boldsymbol{A}_i u_i$ is a linear combination of the PC coefficients. The eigenvalue approximation used here is given by

$$\begin{aligned} \max \lambda_{\boldsymbol{A}} &= \max_{\boldsymbol{v} \in \mathbb{R}^{M+1}} \frac{\boldsymbol{v}^T (\sum \boldsymbol{A}_i u_i) \boldsymbol{v}}{\boldsymbol{v}^T \boldsymbol{v}} \leq \sum_{i=0}^{M} \max_{\boldsymbol{v}_i \in \mathbb{R}^{M+1}} \frac{\boldsymbol{v}_i^T \boldsymbol{A}_i \boldsymbol{v}_i}{\boldsymbol{v}_i^T \boldsymbol{v}_i} |u_i| \\ &= \sum_i |u_i| \max |\lambda_{\boldsymbol{A}_i}| . \end{aligned} \tag{6.33}$$

Since $\boldsymbol{A}_0 = \boldsymbol{I}$, this approximation coincides with the exact eigenvalues for a boundary value with $u_i = 0$ for $i > 1$. This can be seen by observing that if $\boldsymbol{x}_1$ is an eigenvector with corresponding eigenvalue λ for the matrix $\boldsymbol{A}_1$, then $\boldsymbol{A}_1 \boldsymbol{x}_1 = \lambda \boldsymbol{x}_1$ and

$$(\boldsymbol{A}_1 u_1 + \boldsymbol{A}_0 u_0)\boldsymbol{x}_1 = u_1 \lambda \boldsymbol{x}_1 + u_0 \boldsymbol{I} \boldsymbol{x}_1 = (u_1 \lambda + u_0)\boldsymbol{x}_1, \tag{6.34}$$

so $u_1 \lambda + u_0$ and $\boldsymbol{x}_1$ are an eigenvalue-eigenvector pair of the matrix $\boldsymbol{A} = \boldsymbol{A}_0 u_0 + \boldsymbol{A}_1 u_1$. This shows that (6.28) is an appropriate eigenvalue approximation for problems where only u_0 and u_1 are nonzero on the boundaries.

For a given boundary condition, the maximum eigenvalue of $\boldsymbol{A}_0$ corresponding to the deterministic part of the condition does not change with increasing number of PC coefficients. However, the largest eigenvalue contribution from $\boldsymbol{A}_1$ grows with the number of PC coefficients.

The eigenvalue approximation (6.28) is in general of the same order of magnitude as the largest eigenvalue in the interior of the domain but might have to be adjusted to remove all oscillations. The exact value is problem specific, and an estimate based on the interior values requires knowledge about the solution to the problem.

6.7 Efficiency of the Polynomial Chaos Method

The convergence of the polynomial chaos expansion is investigated by measuring the discrete Euclidean error norm of the variance and the expected value. For a discretization with m spatial grid points, we have

$$\left\|\epsilon_{Exp}\right\|^2 = \frac{1}{m-1}\sum_{i=1}^{m}(\mathrm{E}[u]_i - \mathrm{E}[u_{ref}]_i)^2$$

and

$$\|\epsilon_{Var}\|^2 = \frac{1}{m-1}\sum_{i=1}^{m}(\mathrm{Var}[u]_i - \mathrm{Var}[u_{ref}]_i)^2$$

where u_{ref} denotes the analytical solution. Consider the model problem (6.9); the problem is solved with the Monte Carlo method and gPC until time $t = 0.3$. Accuracy (measured as the norm of the difference between the actual solution and the analytical solution) and simulation cost are shown in Table 6.1 for the Monte Carlo method and Table 6.2 for the gPC expansions.

For this highly non-linear and discontinuous problem, the polynomial chaos method is more efficient than the Monte Carlo method with low accuracy requirements. The convergence properties of these solutions are affected by the spatial grid size and the accuracy of imposed artificial dissipation and no general conclusion of the relative performances of the two methods will be drawn here. As will be further illustrated in the section on analysis of characteristics, the solution coefficients of the truncated system are discontinuous approximations to the analytical coefficients which are smooth. Even though the gPC results do converge for this problem, the low-order expansions are qualitatively very different from the analytical solution, see for instance Fig. 6.4. Also, note that excessive use of artificial dissipation might

Table 6.1 Convergence to (6.12) and (6.13) with the Monte Carlo method, $m = 400$, $t = 0.3$

N	10	50	100	400	1,600
$\left\|\epsilon_{Exp}\right\|$	0.122	0.0374	0.0344	0.0257	0.0151
$\|\epsilon_{Var}\|$	0.127	0.0589	0.0426	0.0283	0.0189
T (s)	240	1,180	2,390	9,350	38,460

Table 6.2 Convergence to (6.12) and (6.13) with the polynomial chaos method, $m = 400$, $t = 0.3$

M	2	4	6	8
$\left\|\epsilon_{Exp}\right\|$	0.113	0.0544	0.0164	0.0150
$\|\epsilon_{Var}\|$	0.147	0.122	0.0409	0.0630
T (s)	126	636	4,180	10,900

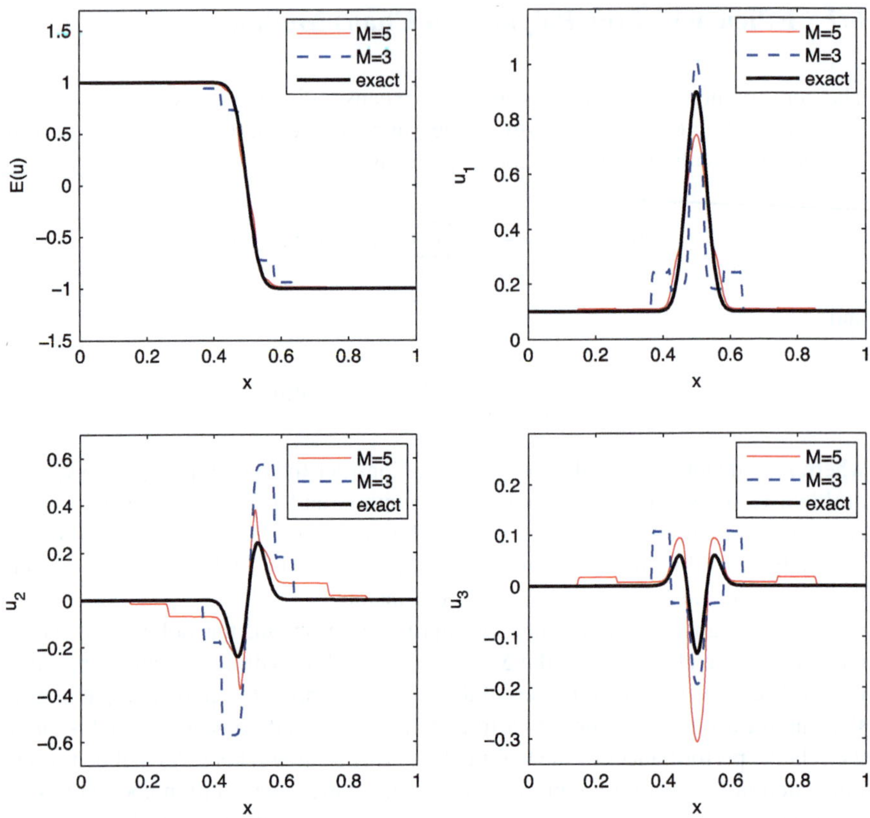

Fig. 6.4 The first four gPC coefficients, $t = 0.3$, $M = 5$ and $M = 3$, $m = 400$

produce an erroneous solution which appears to be closer to the analytical solution for lower-order expansions. Note that, as expected, spatial grid refinement leads to convergence to the true solution of the truncated system but does not get any closer to the analytical solution.

The use of artificial dissipation proportional to the largest eigenvalue makes the solutions of high-order expansions dissipative and spatial grid refinement is needed for an accurate solution. This can be seen in Table 6.2, where the accuracy of the variance decreases with large M.

6.7.1 Numerical Convergence

The convergence of the computed polynomial chaos coefficients, the expected value, and the variance of the truncated system are investigated and comparisons to the analytical solution derived in Sect. 6.2 are presented.

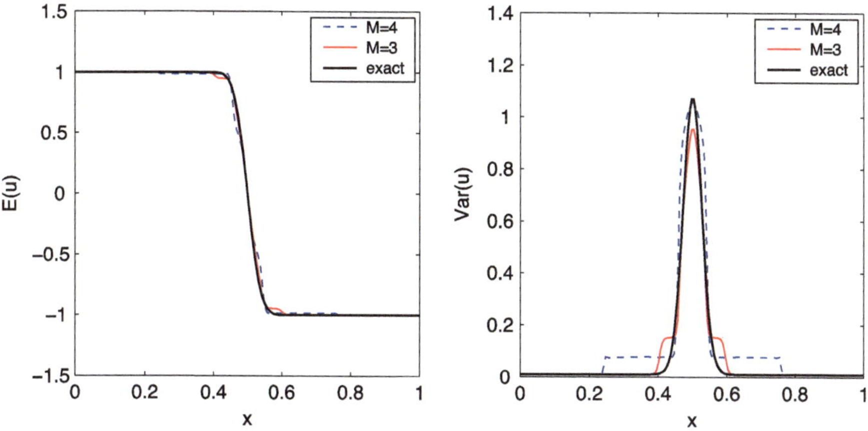

Fig. 6.5 Dissipative solution on coarse grid ($m = 200$), computed for $M = 3$ and non-dissipative solution for $M = 4$

Table 6.3 Norms of errors for dissipative and non-dissipative solutions

M	3	3 (dissipative)	4
$\|\epsilon_{Exp}\|$	0.0354	0.0173	0.0374
$\|\epsilon_{Var}\|$	0.0918	0.0370	0.0723

As mentioned earlier, the numerical results obtained for a small number of expansion terms are expected to be a poor approximation to the analytical solution, as confirmed by the mesh refinement study reported in Fig. 6.4 for $M = 5$. In this particular application, the analytical solution admits continuous (smooth) coefficients in spite of the discontinuous initial condition; on the other hand, the coefficients of the truncated system are discontinuous.

Interestingly, the difference between the computed coefficients corresponding to a finite gPC expansion (u_i for $i \leq M$) and the analytical ($M = \infty$) coefficients indicates that a poorly resolved numerical solution with excessive dissipation might be qualitatively closer to the analytical solution than a grid-converged solution to the truncated system. Figure 6.5 and Table 6.3 illustrate this phenomenon of illusory convergence.

The discrepancy between the truncated solution for $M = 3$ and the analytical solution is also illustrated in Fig. 6.6. The coefficients do not converge to the analytical solution when the spatial grid is refined (Fig. 6.6a, left). Instead, the coefficients converge numerically to a reference solution corresponding to a numerical solution obtained with a large number of gridpoints (Fig. 6.6a, right). For the seventh-order expansion, the solution is sufficiently close to the solution of the analytical problem to exhibit spatial numerical convergence of the first four coefficients to the values of the analytical coefficients (Fig. 6.6b).

The variance calculated for $M = 7$ appears to converge to a function that is close but not equal to the analytical variance given by (6.13) (see Fig. 6.7).

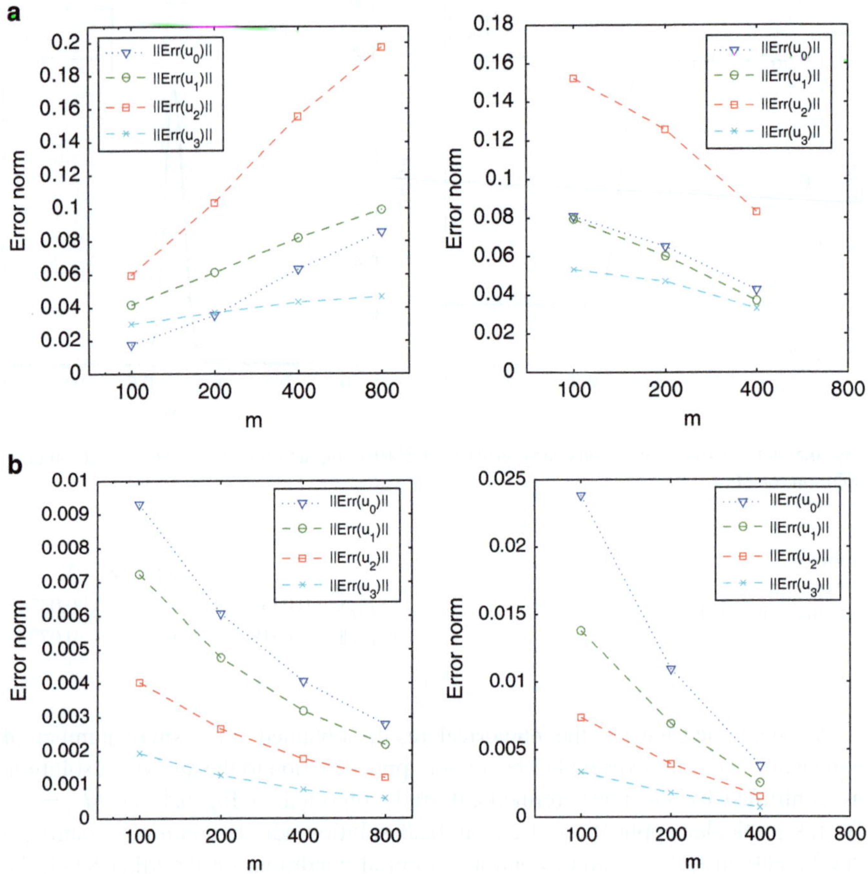

Fig. 6.6 Convergence of the first chaos coefficients. Note the different scales in the figures. (**a**) $M = 3$. Norm of the error relative to the analytical solution (*left*) and error relative to the finest grid solution, $m = 800$ (*right*). (**b**) $M = 7$. Norm of the error relative to the analytical solution (*left*) and error relative to the finest grid solution, $m = 800$ (*right*)

6.8 Theoretical Results and Interpretation

6.8.1 Analysis of Characteristics: Disturbed Cosine Wave

In this section, the characteristics of the stochastic Burgers' equation with $M = 1$ (truncated to 2×2 system) will be investigated to give a qualitative measure of the time development of the solution. The system is given by

$$\begin{pmatrix} u_0 \\ u_1 \end{pmatrix}_t + \begin{pmatrix} u_0 & u_1 \\ u_1 & u_0 \end{pmatrix} \begin{pmatrix} u_0 \\ u_1 \end{pmatrix}_x = \begin{pmatrix} 0 \\ 0 \end{pmatrix}. \tag{6.35}$$

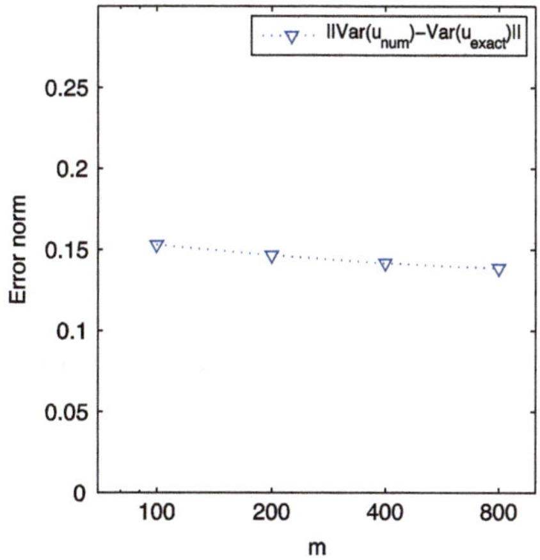

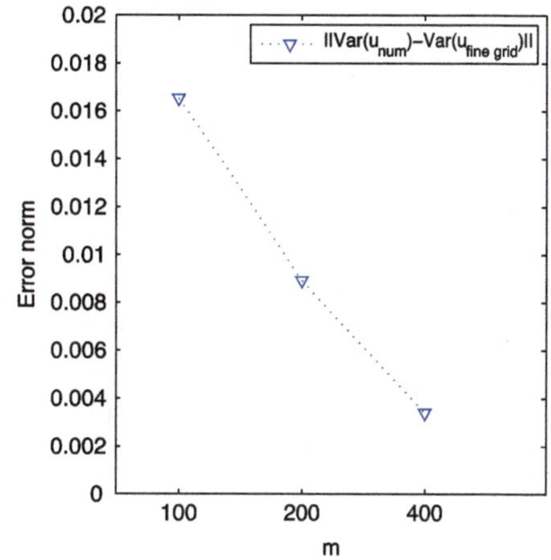

Fig. 6.7 $M = 7$. Convergence of the variance. Norm of the error relative to the analytical variance (*left*) and error relative to the finest grid variance, $m = 800$ (*right*)

With $w_1 = u_0 + u_1$ and $w_2 = u_0 - u_1$, (6.35) can be diagonalized and rewritten

$$\begin{pmatrix} w_1 \\ w_2 \end{pmatrix}_t + \begin{pmatrix} w_1 & 0 \\ 0 & w_2 \end{pmatrix} \begin{pmatrix} w_1 \\ w_2 \end{pmatrix}_x = \begin{pmatrix} 0 \\ 0 \end{pmatrix}. \tag{6.36}$$

Equation (6.36) is the original Burgers' equation for w_1, w_2 and the shock speeds are given by

$$s_{w_1} = \frac{[f(w_1)]}{[w_1]} = \frac{w_{1R} + w_{1L}}{2} = \bar{u}_0 + \bar{u}_1 \tag{6.37}$$

and

$$s_{w_2} = \frac{[f(w_2)]}{[w_2]} = \frac{w_{2R} + w_{2L}}{2} = \bar{u}_0 - \bar{u}_1, \tag{6.38}$$

where we have introduced the mean over the shock, $\bar{u}_i = (u_{iL} + u_{iR})/2$. Square brackets [ ] denote the jump in a quantity over a discontinuity. Similarly to (6.37) and (6.38), with the non-diagonalized system in conservation form, the propagation speeds of discontinuities in u_0, u_1 are given by

$$s_{u_0} = \frac{[(u_0^2 + u_1^2)/2]}{[u_0]} = \bar{u}_0 + \bar{u}_1 \frac{[u_1]}{[u_0]} \tag{6.39}$$

and

$$s_{u_1} = \frac{[u_0 u_1]}{[u_1]} = \bar{u}_0 + \bar{u}_1 \frac{[u_0]}{[u_1]}. \tag{6.40}$$

The analysis of characteristics w_1 and w_2 describes the behavior and emergence of discontinuities in the coefficients u_0 and u_1 of the truncated system. However, the coefficients of the solution to the problem given by the infinite gPC expansion are smooth (except for $t = 0$ for the Riemann problem). Diagonalization of large systems is not feasible, but we can obtain expressions for the shock speeds of the coefficients. For instance, the expression (6.39) for the shock speed in u_0 can be generalized for gPC expansions of order M as

$$s_{u_0} = \sum_{i=0}^{M} \bar{u}_i \frac{[u_i]}{[u_0]}. \tag{6.41}$$

In the assumption that only one Gaussian variable ξ is introduced, and the uncertainty is (linearly) proportional to ξ, only a limited number of different values of the correlation coefficient between the left and right states can occur. Since we are also assuming the same model for the left and right state uncertainties, only a few combinations of covariance matrices describing their correlation are realizable. With the assumptions made here, the dependence between the two states is determined by the correlation coefficient ρ_{LR}, which for these cases is either 1 or -1.

Ex 1.1

$$u(x,0,\xi) = \begin{cases} u_L = 1 + \hat{\sigma}\xi & x < x_0 \\ u_R = -1 - \hat{\sigma}\xi & x < x_0 \end{cases}$$

$$u(x,0) = cos(\pi x)(1 + \hat{\sigma}\xi)$$

$$\xi \sim \mathcal{N}(0,1)$$

$$\rho_{LR} = -1.$$

Ex 1.2

$$u(x,0,\xi) = \begin{cases} u_L = 1 + \hat{\sigma}\xi & x < x_0 \\ u_R = -1 + \hat{\sigma}\xi & x < x_0 \end{cases}$$

$$u(x,0) = cos(\pi x) + \hat{\sigma}\xi$$

$$\xi \sim \mathcal{N}(0,1)$$

$$\rho_{LR} = 1.$$

The problems are similar in terms of expected value and variance at the boundary, but the difference in correlation between the left and right states completely changes the behavior over time. The difference in initial variance in the interior of the domain has only a limited impact on the time-dependent difference between the solutions; this has been checked by varying the initial functions. Note that Ex 1.1 is included to show the importance of the sign of the stochastic variable, but it is a special case of a more general phenomenon of superimposition of discontinuities exhibited by Ex 1.2 and further explained and analyzed below. Figure 6.8 shows the two cases at time $t = 0.5$ for $M = 3$. We use $\hat{\sigma} = 0.1$ and $x_0 = 0.5$.

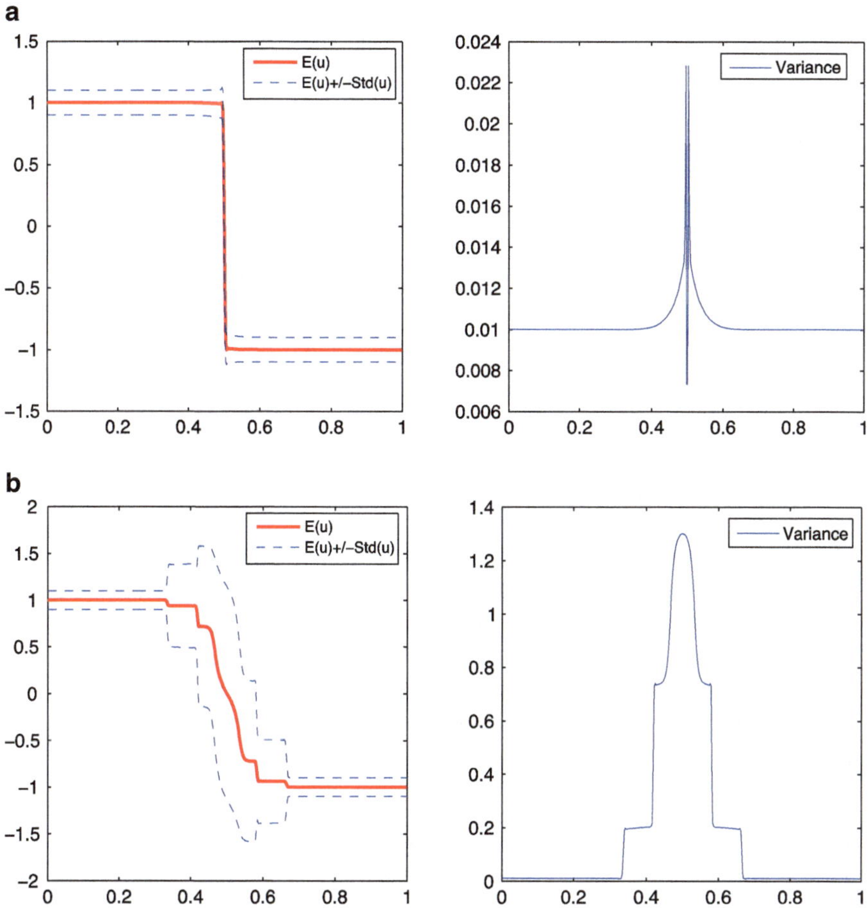

Fig. 6.8 Development of variance of the perturbed cosine wave. $t = 0.5$ for $M = 3$, $m = 400$. (**a**) Ex 1.1. Symmetric boundary conditions. (**b**) Ex 1.2. Constant initial variance

To explain the differences between the solutions depicted in Fig. 6.8, we turn to analysis of the characteristics for the truncated system with $M = 1$. The polynomial chaos coefficients of the boundaries are given by

$$\left.\begin{array}{l} u_0 = 1 \\ u_1 = 0.1 \end{array}\right\} x = 0, \quad \left.\begin{array}{l} u_0 = -1 \\ u_1 = -0.1 \end{array}\right\} x = 1 \quad \text{(Ex 1.1)}$$

and

$$\left.\begin{array}{l} u_0 = 1 \\ u_1 = 0.1 \end{array}\right\} x = 0, \quad \left.\begin{array}{l} u_0 = -1 \\ u_1 = 0.1 \end{array}\right\} x = 1 \quad \text{(Ex 1.2)},$$

respectively.

Note that with more PC coefficients included, the higher-order coefficients are zero at the boundaries for sufficiently short times. The expected boundary values as well as the boundary variance are the same for Ex 1.1 and Ex 1.2. In order to relate the concepts of characteristics with expected value and variance, we will use the fact that the expected value at each point is the average of the characteristics,

$$\mathrm{E}(u) = u_0 = \frac{w_1 + w_2}{2}, \tag{6.42}$$

and that the variance depends on the distance between the characteristics,

$$\mathrm{Var}(u) = u_1^2 = \left(\frac{w_1 - w_2}{2}\right)^2. \tag{6.43}$$

To explain the qualitative differences between the two cases Ex 1.1 and Ex 1.2, consider the decoupled system (6.36). The boundary values for u_0 and u_1 are inserted into the characteristic variables w_1 and w_2; discontinuities emerge when the characteristics meet.

For Ex 1.1 we have $w_1(x = 0) = -w_1(x = 1)$ and $w_2(x = 0) = -w_2(x = 1)$. Inserting these values in (6.37) and (6.38) gives the shock speeds $s_{w_1} = s_{w_2} = 0$, corresponding to two stationary shocks (of different magnitude) at $x = 0.5$, which can be seen in Fig. 6.9a. Inserting the characteristic values (can be evaluated in Fig. 6.9) into (6.43) results in uniform variance except around the discontinuity, Fig. 6.10a. Since the characteristic solution is propagating from the boundaries, this interval shrinks with time and collapses at $x = 0.5$.

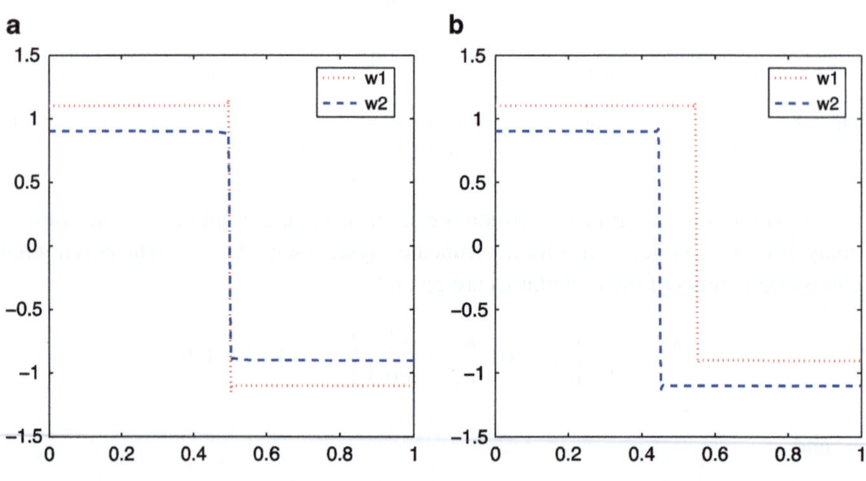

Fig. 6.9 Characteristics of the two perturbed cosine waves (Ex 1.1 and Ex 1.2) for $M = 1$. **(a)** Ex 1.1. The variance is undefined at $x = 0.5$. **(b)** Ex 1.2. The variance peaks at $x = 0.5$. w_1 is left-going and w_2 is right-going

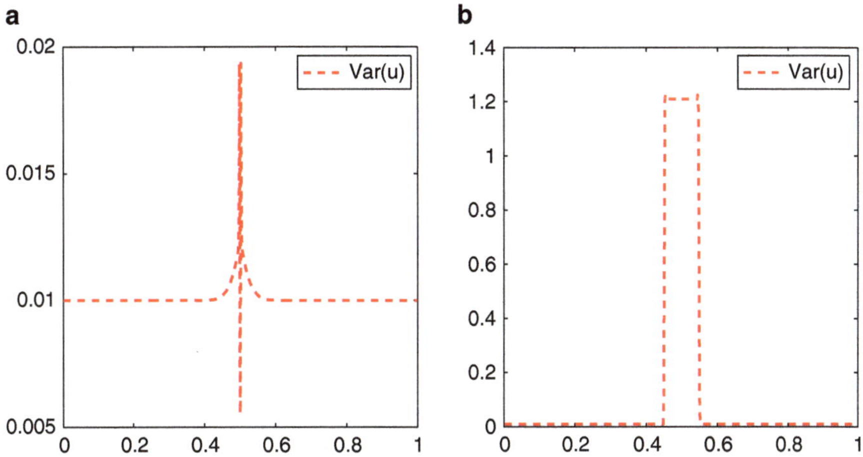

Fig. 6.10 Variance of Ex 1.1 and Ex 1.2 for $M = 1$, calculated from w_1, w_2 using (6.43). (**a**) Ex 1.1. The variance is constant except around the discontinuity. (**b**) Ex 1.2. The variance is maximal at the shock location and spreads towards the boundaries

In Ex 1.2, the characteristics are $w_1(x = 0) = 1.1 > -w_1(x = 1) = 0.9$ and $w_2(x = 0) = 0.9 < -w_2(x = 1) = 1.1$. Evaluating (6.37) and (6.38) when the characteristics cross yields $s_{w_1} = 0.1$ and $s_{w_2} = -0.1$. When the characteristics meet, the discontinuity will split and propagate as two moving shocks in u_0 and u_1, located equidistantly from the midpoint $x = 0.5$. In w_1 and w_2 there will still be a single shock. The shock speeds are given by the expressions (6.37)–(6.40). The vertical gap between the characteristics at $x = 0.5$ in Fig. 6.9b corresponds to the variance peak at this location in Fig. 6.8b.

The system used for analysis of characteristics is truncated to $M = 1$, but the conclusions about the qualitative behavior holds for higher-order systems. Including more polynomial chaos coefficients would result in additional shocks of different magnitude and speed. Observe the qualitative similarities between the solutions in Figs. 6.8 and 6.9. Regardless of the truncation of PC coefficients, the variance approaches 0 at the shock location in Ex 1.1. At the shock location in Ex 1.2, the variance reaches a maximum that will spread towards the boundaries and cancel the discontinuity. The observation that the same boundary and initial expected value and variance can give totally different solutions indicates that knowledge about the PC coefficients is required to obtain a unique solution.

Further analysis shows that the problem could be partitioned into several phases of development, depending on the speeds of the characteristics. Consider again the boundary conditions of Ex 1.1 and Ex 1.2 but now assume $u(x, 0) = 0$ for $x \in$

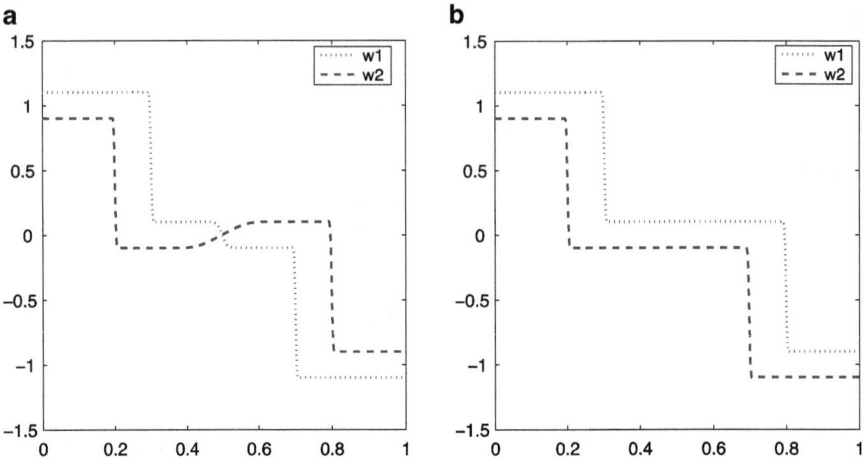

Fig. 6.11 Characteristics at $t = 0.5$, $M = 1$. (**a**) Ex 1.1. Boundary conditions: $u(0,t) = (1, 0.1, 0, \ldots)$; $u(1,t) = (-1, -0.1, 0, \ldots)$. (**b**) Ex 1.2. Boundary conditions: $u(0,t) = (1, 0.1, 0, \ldots)$; $u(1,t) = (-1, 0.1, 0, \ldots)$

$(0, 1)$. The solution for $M = 1$ before the characteristics meet is shown in Fig. 6.11. With more PC coefficients, the sharp edges in the solution will disappear. At time $t = 0.5$, the solutions to the two problems are still similar, with two variance peaks at the shocks that are traveling towards the middle of the domain. For comparison, Fig. 6.12 shows the expected value and variance calculated from the characteristics in Fig. 6.11.

Asymptotically in time, the symmetric problem (Ex 1.1) will result in a stationary shock. The variance will equal the initial boundary variance except for a peak at the very location of the shock. The boundary conditions are independent of time. This property is illustrated in Fig. 6.13a, where the solution has reached steady-state.

The time development of the solution of Ex 1.2 is not consistent with the stationary boundary conditions stated in the problem formulation. The characteristics are transported from one boundary to the other (see Fig. 6.13b), thus changing the boundary data. The boundary conditions of Ex 1.2 must therefore be time-dependent (and can be calculated exactly from (6.10) for this example). Unlike the continuously varying boundary conditions of the full PC expansion problem, the boundary conditions for the truncated system of Fig. 6.13b will change discontinuously from the initial boundary condition to zero at the moment the characteristics reach the boundaries. In a general hyperbolic problem, the imposition of correct time-dependent boundary conditions might become one of the more significant problems with the gPC method. A detailed investigation is necessary to

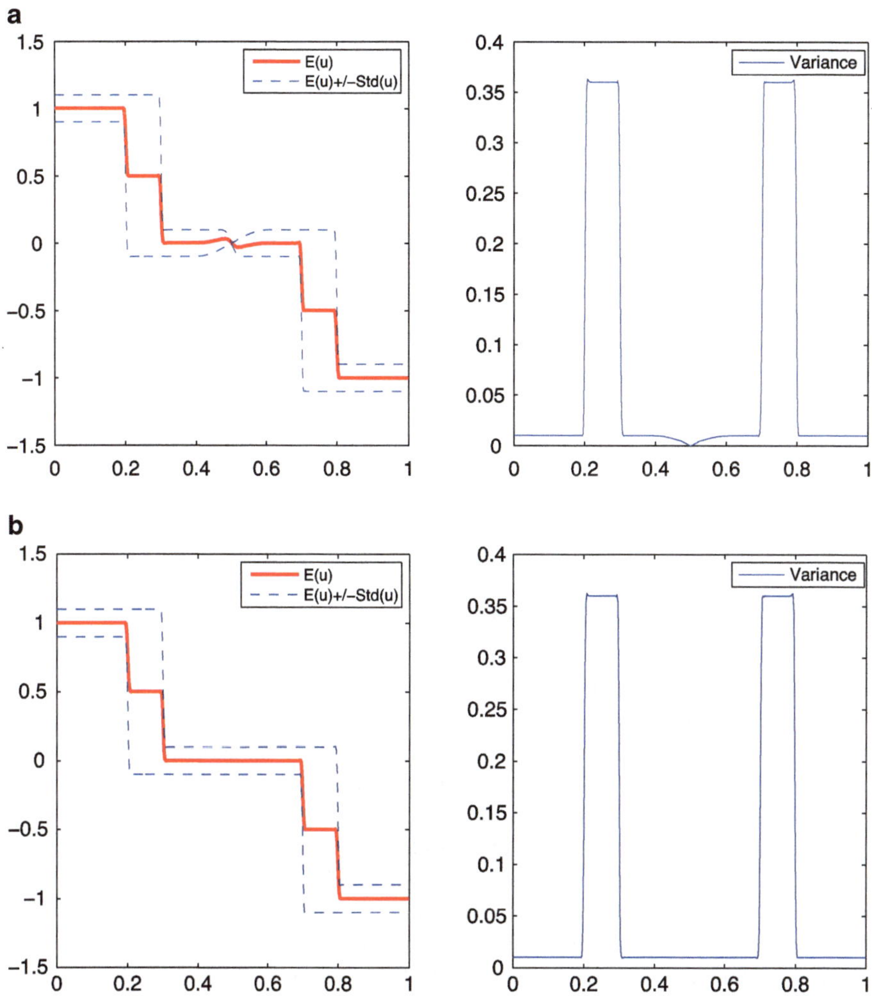

Fig. 6.12 Expected value and variance at $t = 0.5$, $M = 1$. (**a**) Ex 1.1. Symmetric boundary conditions: $u(0,t) = (1, 0.1, 0, \ldots)$; $u(1,t) = (-1, -0.1, 0, \ldots)$. (**b**) Ex 1.2. Boundary conditions: $u(0,t) = (1, 0.1, 0, \ldots)$; $u(1,t) = (-1, 0.1, 0, \ldots)$

identify an approach to specify time-dependent stochastic boundary data, especially for the higher-order moments. Special non-reflecting boundary conditions will be required. In the case studied here, analytical boundary conditions have been derived and can be correctly imposed for any time and any order of PC expansion.

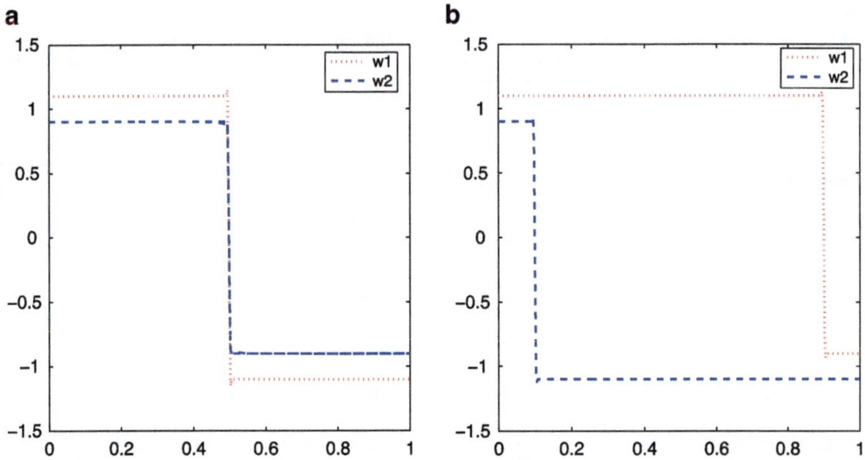

Fig. 6.13 Characteristics at $t = 4$ for $M = 1$. (**a**) Ex 1.1. Characteristics have reached steady-state. (**b**) Ex 1.2. w_1 is right-going, w_2 left-going

6.9 Summary and Conclusions

The stochastic Galerkin method has been presented for Burgers' equation with uncertain boundary conditions. Stable difference schemes are obtained by the use of artificial dissipation, difference operators satisfying the summation by parts property and a weak imposition of characteristic boundary conditions.

A number of mathematical properties of the deterministic Burgers' equation hold for the hyperbolic problem that results from the Galerkin projection of the truncated PC expansions. The system is symmetric, and a split form combining conservative and non-conservative formulations is used to obtain an energy estimate. The truncated linearized problem is shown to be well-posed. The system eigenvalues vary over time and this makes the choice of the time-step difficult; moreover, this affects the accuracy of the methods since the dissipation operators are eigenvalue dependent. An eigenvalue estimate is provided.

To devise a suitable numerical method, we need to know whether the solution we seek is smooth or discontinuous. Even though the solution to the Burgers' equation is discontinuous for a particular value of the uncertain (stochastic) variable, the PC coefficient functions are in general continuous for the Riemann problems investigated. The solution coefficients of the truncated system are discontinuous and can be treated as a superimposition of a finite number of discontinuous characteristic variables. This has been shown explicitly for the 2×2-case. The discontinuous coefficients converge with the number of PC coefficients to continuous functions.

Examples have shown the need to provide time-dependent boundary conditions that might include higher-order moments. Stochastic time-dependent boundary conditions have been derived for the Burgers' equation.

An increasing number of polynomial chaos modes and use of extra boundary data give solutions that are qualitatively different from the cruder approximation. However useful for a qualitative description of the dynamics of the hyperbolic system, the approximation error due to truncation of the infinite polynomial chaos series is dominating the total error.

As shown in Table 6.3, excessive use of artificial dissipation can give a numerical solution that more closely resembles the solution to the original problem compared to a solution where a small amount of dissipation (within the order of accuracy) is used to preserve the discontinuities of the truncated solution. Clearly, only the latter method could be justified from a theoretical point of view.

In general, excessively dissipative schemes should be avoided and, if possible, mesh refinement studies should be performed to ensure numerical convergence. In addition to the problems associated with non-converged deterministic solutions, a non-converged stochastic solution is likely to misrepresent the variance and other higher-order statistics.

6.10 Supplementary Codes

Matlab scripts for the Burgers' equation with the stochastic Galerkin method and SBP operators are provided at [http://extras.springer.com]. The reader is encouraged to experiment with the scripts and use them as a complement to the exercises.

To get started with the codes, simply run the main script `burgers_main.m` in Matlab. The scheme is not conditionally stable, so changes may lead to numerical instability unless the parameters follow the derivations of this chapter. The script `SBP_operators.m` contains SBP operators of orders 2,4,6 and 8 and is a generic implementation that may be used for other problems of interest. Note that in the context of hyperbolic problems, the artificial dissipation operators should be adjusted to the order of SBP operators.

For the problem of interest, the analytical solution of the original problem (the gPC coefficients of the infinite order expansion) is known, as is the exact solution of the truncated $M = 1$ problem. Depending on the order of truncation, one may compare the numerical solution to the infinite order case, or to the first-order expansion.

6.11 Exercises

6.1. Consider stochastic Galerkin projection of the Riemann problem (6.9) of some finite order M using normalized Hermite polynomials. As you vary b (standard deviation) in relation to a (expectation), how does the number of boundary conditions change?

6.2. Consider the Riemann problem introduced in Sect. 6.2 but with an inverted initial condition

$$
\begin{aligned}
&u(x,0,\xi) = \begin{cases} u_L = -a + p(\xi) & \text{if } x < x_0 \\ u_R = a + p(\xi) & \text{if } x > x_0 \end{cases} \\
&u(0,t,\xi) = u_L,\ u(1,t,\xi) = u_R \\
&\xi \in \mathcal{N}(0,1).
\end{aligned}
\tag{6.44}
$$

Compare the solution (in terms of gPC modes) of the truncated vs. full expansion system at t=0.1 and t=0.5.

6.3. In Sect. 6.8.1, two examples are introduced to show the effect of boundary conditions. Repeat the analysis using the following inverted initial conditions:

Ex 1.1, modified

$$u(x,0,\xi) = \begin{cases} u_L = -1 + \hat{\sigma}\xi & x < x_0 \\ u_R = 1 - \hat{\sigma}\xi & x < x_0 \end{cases}$$

$$u(x,0) = cos(\pi x)(-1 + \hat{\sigma}\xi)$$

$$\xi \sim \mathcal{N}(0,1)$$

$$\rho_{LR} = -1.$$

Ex 1.2, modified

$$u(x,0,\xi) = \begin{cases} u_L = -1 + \hat{\sigma}\xi & x < x_0 \\ u_R = 1 + \hat{\sigma}\xi & x < x_0 \end{cases}$$

$$u(x,0) = -cos(\pi x) + \hat{\sigma}\xi$$

$$\xi \sim \mathcal{N}(0,1)$$

$$\rho_{LR} = 1.$$

6.4. Consider the initial condition $u(x,0,\xi) = \alpha \tanh \frac{x-0.5}{0.06} + 3.0\xi \exp\left(-\frac{(x-0.5)^2}{0.045}\right)$ with ξ a normal random variable with zero mean and unit variance in the domain $x \in [-3:3]$. Study the convergence of the Galerkin expansion for $\alpha = \pm 1$.

6.5. Consider the same problem for a slightly different initial condition $u(x,0,\xi) = \alpha \tanh \frac{x-0.5}{0.06} + 3.0\xi \exp\left(-\frac{(x-0.5)^2}{0.045}\right) \sin(5.3\pi x)$.

References

1. Carpenter MH, Gottlieb D, Abarbanel S (1994) Time-stable boundary conditions for finite-difference schemes solving hyperbolic systems: methodology and application to high-order compact schemes. J Comput Phys 111(2):220–236. doi:http://dx.doi.org/10.1006/jcph.1994.1057
2. Carpenter MH, Nordström J, Gottlieb D (1999) A stable and conservative interface treatment of arbitrary spatial accuracy. J Comput Phys 148(2):341–365. doi:http://dx.doi.org/10.1006/jcph.1998.6114
3. Chen QY, Gottlieb D, Hesthaven JS (2005) Uncertainty analysis for the steady-state flows in a dual throat nozzle. J Comput Phys 204(1):378–398. doi:http://dx.doi.org/10.1016/j.jcp.2004.10.019
4. Gottlieb D, Hesthaven JS (2001) Spectral methods for hyperbolic problems. J Comput Appl Math 128(1–2):83–131. doi:http://dx.doi.org/10.1016/S0377-0427(00)00510-0

5. Gottlieb D, Xiu D (2008) Galerkin method for wave equations with uncertain coefficients. Commun Comput Phys 3(2):505–518
6. Hou TY, Luo W, Rozovskii B, Zhou HM (2006) Wiener chaos expansions and numerical solutions of randomly forced equations of fluid mechanics. J Comput Phys 216(2):687–706. doi:http://dx.doi.org/10.1016/j.jcp.2006.01.008
7. LeVeque RJ (2006) Numerical methods for conservation laws, 2nd edn. Birkhäuser Basel
8. Mattsson K, Svärd M, Nordström J (2004) Stable and accurate artificial dissipation. J Sci Comput 21(1):57–79
9. Nordström J (2006) Conservative finite difference formulations, variable coefficients, energy estimates and artificial dissipation. J Sci Comput 29(3):375–404. doi:http://dx.doi.org/10.1007/s10915-005-9013-4
10. Nordström J, Carpenter MH (1999) Boundary and interface conditions for high-order finite-difference methods applied to the Euler and Navier-Stokes equations. J Comput Phys 148(2):621–645. doi:http://dx.doi.org/10.1006/jcph.1998.6133
11. Nordström J, Carpenter MH (2001) High-order finite difference methods, multidimensional linear problems, and curvilinear coordinates. J Comput Phys 173(1):149–174. doi:http://dx.doi.org/10.1006/jcph.2001.6864
12. Pettersson P, Iaccarino G, Nordström J (2009) Numerical analysis of the Burgers' equation in the presence of uncertainty. J Comput Phys 228:8394–8412. doi:10.1016/j.jcp.2009.08.012, http://dl.acm.org/citation.cfm?id=1621150.1621394
13. Richtmyer RD, Morton KW (1967) Difference methods for initial-value problems, 1st edn. Interscience, New York
14. Xiu D, Karniadakis GE (2003) Modeling uncertainty in flow simulations via generalized polynomial chaos. J Comput Phys 187(1):137–167. doi:10.1016/S0021-9991(03)00092-5, http://dx.doi.org/10.1016/S0021-9991(03)00092-5
15. Xiu D, Karniadakis GE (2004) Supersensitivity due to uncertain boundary conditions. Int J Numer Meth Eng 61:2114–2138

Chapter 7
Boundary Conditions and Data

In this chapter, based on the work in [1], we continue analysis of Burgers' equation with a focus on the effect of data for the boundary conditions. To facilitate understanding, we deal only with the truncated representation $u(x, t, \xi) = u_0\psi_0 + u_1\psi_1$. This means that all the stochastic variation is accounted for by the single gPC coefficient u_1, and the standard deviation of the solution is simply $|u_1|$. With this simplified setup, we obtain a few combinations of general situations for the boundary data: known expectation but unknown standard deviation, unknown expectation and standard deviation, and so on.

7.1 Dependence on Available Data

For $M = 1$, the system (6.4) can be diagonalized with constant eigenvectors and we get an exact solution to the truncated problem. The solution has two characteristics, moving in directions determined by u_0 and u_1. With a and b as in the problem setup (Sect. 6.2), the analytical solution for the 2×2-system ($x \in [0, 1]$) is given by

$$\begin{pmatrix} u_0 \\ u_1 \end{pmatrix} = \begin{cases} \left.\begin{matrix} (a, b)^T & \text{if } x < x_0 - bt \\ (0, a+b)^T & \text{if } x_0 - bt < x < x_0 + bt \\ (-a, b)^T & \text{if } x > x_0 + bt \end{matrix}\right\} & \text{for } 0 \le t < \frac{x_0}{b} \\ (0, a+b)^T & \text{for } t > \frac{x_0}{b} \end{cases}. \tag{7.1}$$

We expect different numerical solutions depending on the amount of available boundary data. We will assume that the boundary data are known on the boundary $x = 1$ and investigate three different cases for the left boundary $x = 0$ corresponding to a complete set of data, partial information about boundary data and no data available, respectively. For all cases, we will solve the $M = 1$ stochastic Galerkin system of the form

M.P. Pettersson et al., *Polynomial Chaos Methods for Hyperbolic Partial Differential Equations*, Mathematical Engineering,
DOI 10.1007/978-3-319-10714-1_7

$$\begin{pmatrix} u_0 \\ u_1 \end{pmatrix}_t + \frac{1}{2}\left[\begin{pmatrix} u_0 & u_1 \\ u_1 & u_0 \end{pmatrix}\begin{pmatrix} u_0 \\ u_1 \end{pmatrix}\right]_x = 0 \tag{7.2}$$

with boundary data

$$\begin{pmatrix} u_0 \\ u_1 \end{pmatrix}_{x=0} = \begin{pmatrix} g_0(t) \\ g_1(t) \end{pmatrix} ; \quad \begin{pmatrix} u_0 \\ u_1 \end{pmatrix}_{x=1} = \begin{pmatrix} h_0(t) \\ h_1(t) \end{pmatrix}.$$

7.1.1 Complete Set of Data

The boundary conditions are

$$u(0,t) = \begin{cases} (a,b)^T & 0 \le t < \frac{x_0}{b} \\ (0, a+b)^T & t > \frac{x}{b} \end{cases}. \tag{7.3}$$

Consider $a = 1, b = 0.2$. Both u_0 and u_1 are known at $x = 0$ and the two ingoing characteristics are assigned the analytical values. The system satisfies the energy estimate (6.26). Observe that when a full set of data is available, the problem is both strongly well-posed according to (6.18) and strongly stable according to (6.26). Figures 7.1–7.3 show the solution at time $t = 1, t = 2$ and $t = 3$, respectively.

7.1.2 Incomplete Set of Boundary Data

Without a complete set of boundary data, the time-dependent behavior of the solution will be hard to predict. Here we assume that the boundary conditions at

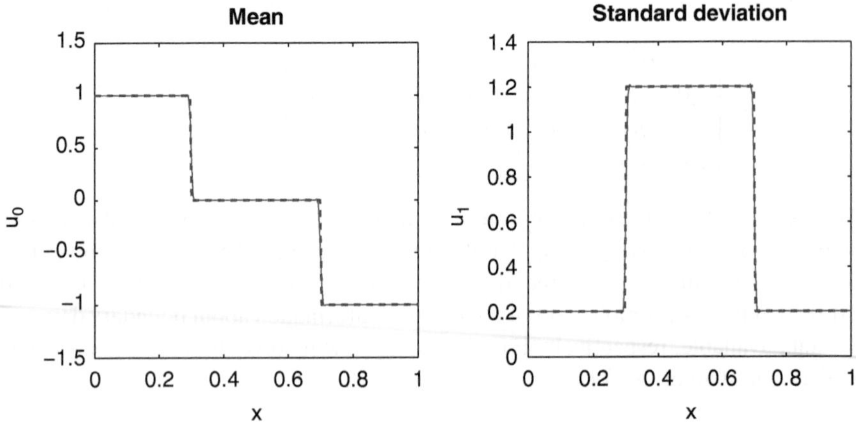

Fig. 7.1 u_0 (*left*) and u_1 (*right*), numerical solution (*solid red*) and exact solution (*dashed blue*). $t = 1$. Complete set of data

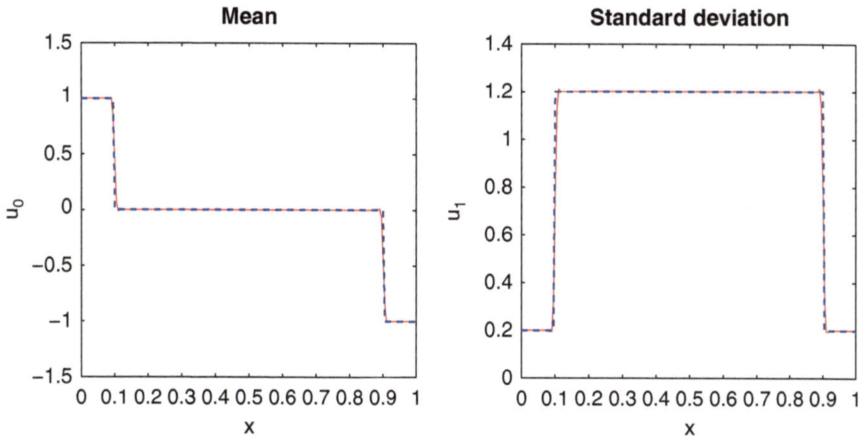

Fig. 7.2 u_0 (*left*) and u_1 (*right*), numerical solution (*solid red*) and exact solution (*dashed blue*). $t = 2$. Complete set of data

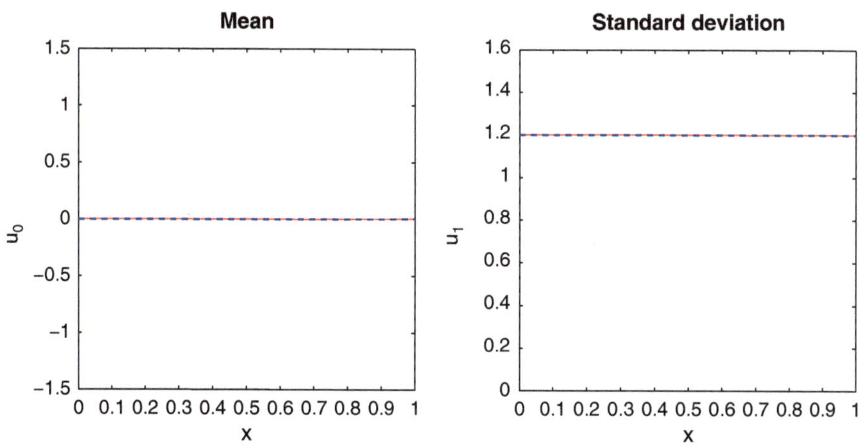

Fig. 7.3 u_0 (*left*) and u_1 (*right*), numerical solution (*solid red*) and exact solution (*dashed blue*). $t = 3$. Complete set of data

$x = 1$ is $u = (-1, 0.2)^T$ as before (7.3) and consider different ways of dealing with unknown data at $x = 0$. Note that with a lack of data, we have different cases depending on how we deal with the situation. The initial function is the same as in the analytical problem above, i.e.,

$$(u_0(x,0), u_1(x,0))^T = \begin{cases} (a,b)^T & \text{if } x < x_0 \\ (-a,b)^T & \text{if } x > x_0 \end{cases}.$$

7.1.2.1 Unknown u_1 at $x = 0$, Guess u_1

When we guess the value of u_1, the continuous problem is strongly well-posed (energy estimate (6.18)) and the semidiscrete problem is strongly stable (energy estimate (6.26)). However, the accuracy of the solution will depend on the guess. First, assume that u_0 is known and u_1 is unknown and put $u_1 = 0.2$ at the boundary for all time. There are two ingoing characteristics at $t = 0$. The value of u_0 at $x = 0$ changes with the boundary conditions of the analytical solution as given by (7.3). The time development of the numerical solution closely follows the analytical solution at first (Fig. 7.4), but eventually becomes inconsistent with the boundary conditions (Figs. 7.5 and 7.6)

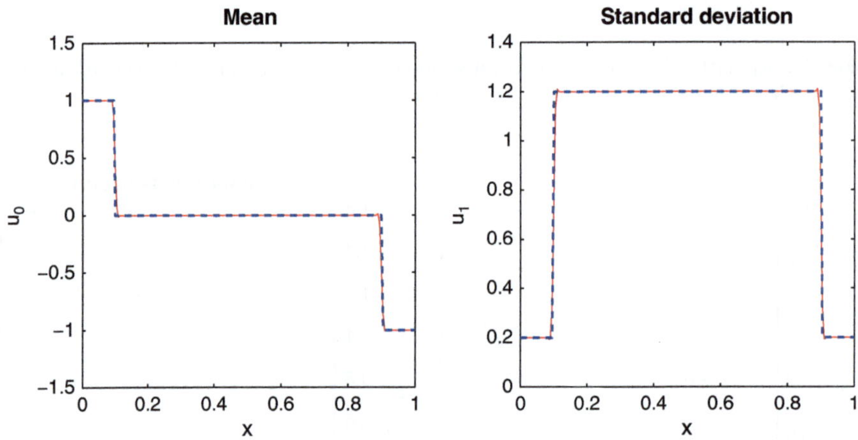

Fig. 7.4 u_1 kept fixed at 0.2, numerical solution (*solid red*) and exact solution (*dashed blue*). $t = 2$

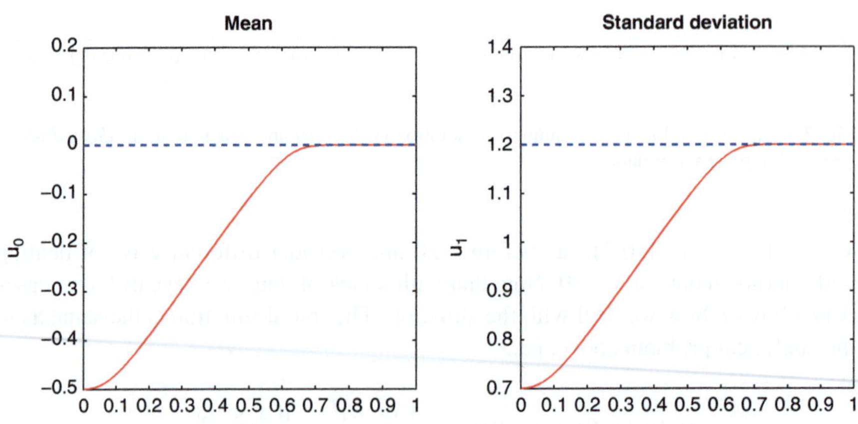

Fig. 7.5 u_1 kept fixed at 0.2, numerical solution (*solid red*) and exact solution (*dashed blue*). $t = 3$

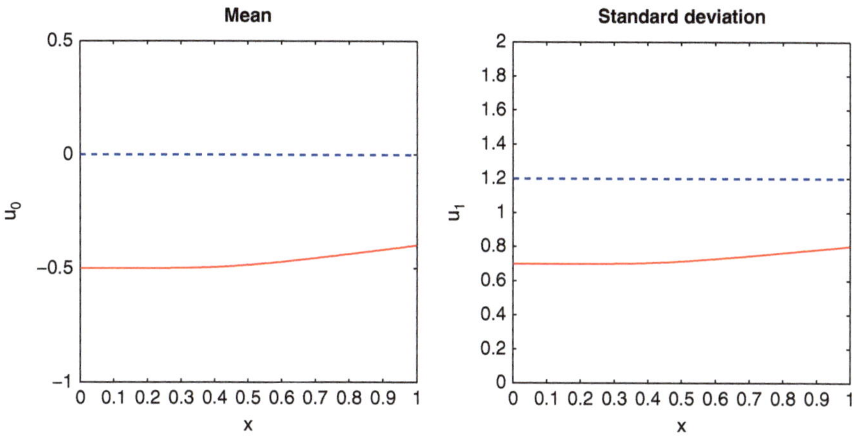

Fig. 7.6 u_1 kept fixed at 0.2, numerical solution (*solid red*) and exact solution (*dashed blue*). $t = 5$

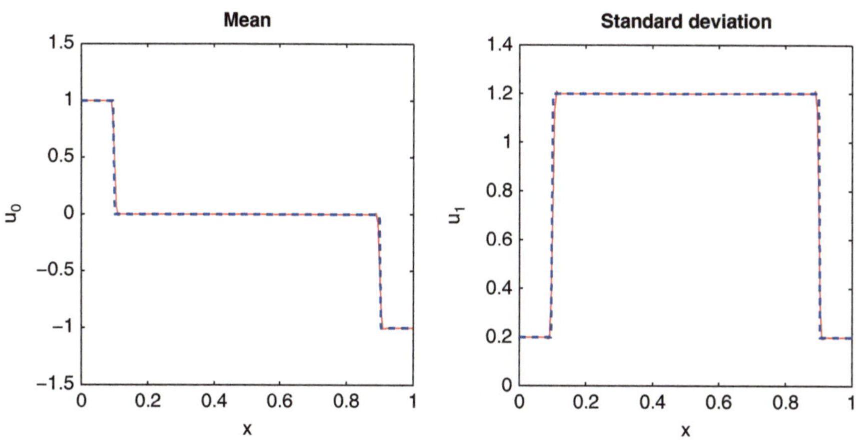

Fig. 7.7 u_1 extrapolated from the interior, numerical solution (*solid red*) and exact solution (*dashed blue*). $t = 2$

7.1.2.2 Unknown u_1 at $x = 0$, Extrapolate u_1

Now, consider the situation where the extrapolation $g_1 = (u_1)_{x=1}$ is used to assign boundary data to the presumably unknown coefficient u_1. When extrapolation is used, we do not impose any data, and the problem is neither well-posed nor stable. The numerical solution does not blow up, but the result is inaccurate. As long as the analytical boundary conditions do not change, the numerical solution will follow the analytical solution as before, see Fig. 7.7. After $t = 2.5$, the characteristics have reached the opposite boundaries and the error grows (Fig. 7.8) before reaching the steady-state solution (Fig. 7.9).

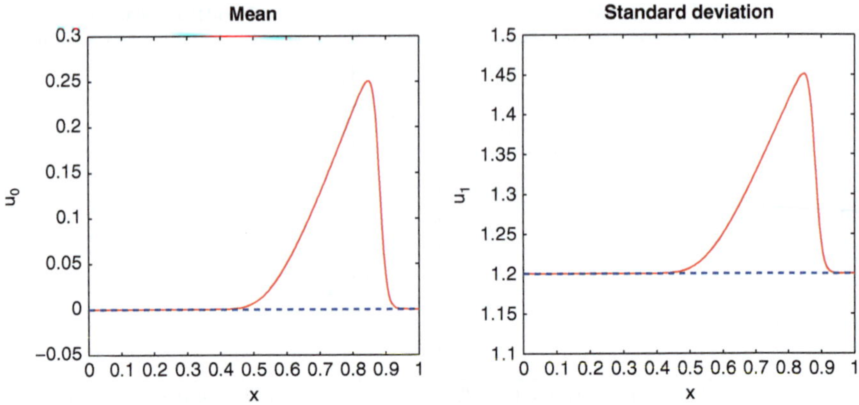

Fig. 7.8 u_1 extrapolated from the interior, numerical solution (*solid red*) and exact solution (*dashed blue*). $t = 3$

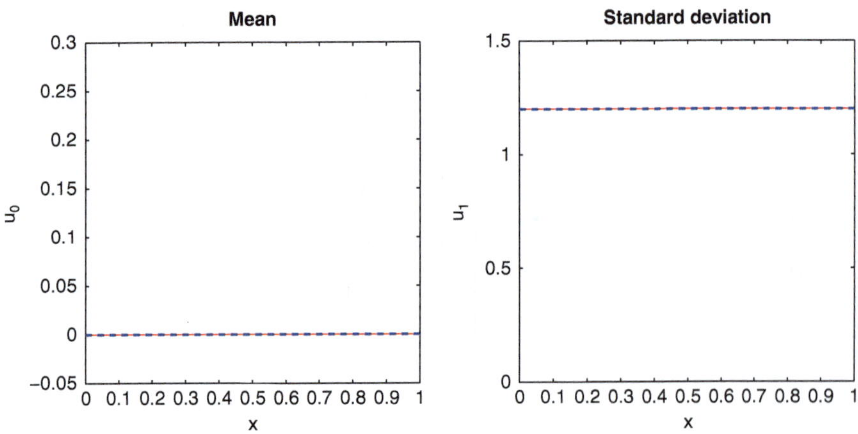

Fig. 7.9 u_1 extrapolated from the interior. $t = 5$, numerical solution (*solid red*) and exact solution (*dashed blue*). The error is of the order 10^{-15}

7.1.2.3 Unknown u_0 at $x = 0$, Guess u_0

Next we assume that the boundary data for u_0 are unknown. With a guessed value of data, the problem is strongly well posed according to the energy estimate (6.18) and strongly stable according to (6.26). However, depending on the guess, the solution can be more or less accurate. The same analysis as was done for u_1 in the preceding section is now carried out for u_0. First, u_0 at $x = 0$ is held fixed for all times. Figures 7.10 and 7.11 show the solution before and after the true characteristics reach the boundaries. Note that the numerical solution after a long time is not

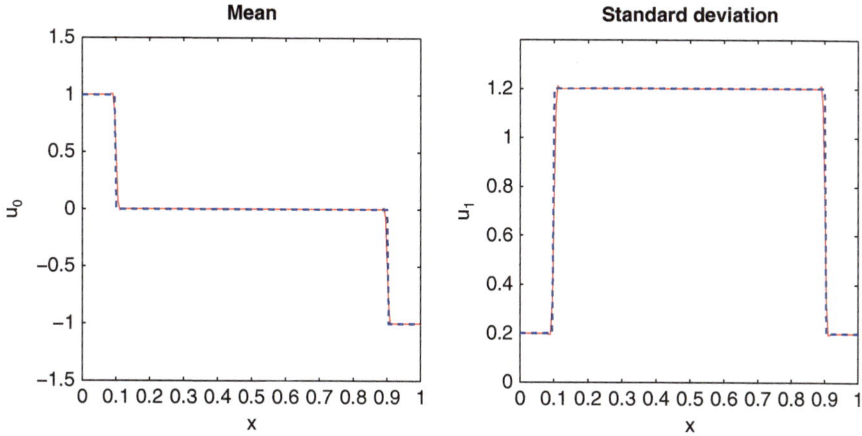

Fig. 7.10 u_0 is held fixed. Numerical solution (*solid red*) and exact solution (*dashed blue*), $t = 2$

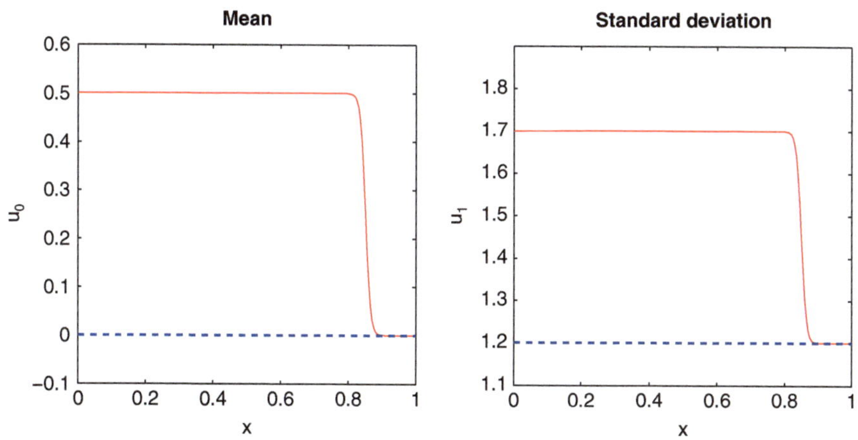

Fig. 7.11 u_0 is held fixed. Numerical solution (*solid red*) and exact solution (*dashed blue*), $t = 3$

coincident with the analytical solution and that the true boundary conditions are not satisfied (Fig. 7.12).

7.1.2.4 Unknown u_0 at $x = 0$, Extrapolate u_0

The data for u_0 can alternatively be extrapolated from the interior of the domain. The extrapolation $g_0 = (u_0)_{x=1}$ is used (see Figs. 7.13–7.15). We do not impose any data when extrapolation from the interior is used, and the problem is neither well-posed nor stable. In this case no explosion occurs, and the result is accurate. Note that the solution after a long time is very close to the analytical solution (Fig. 7.15).

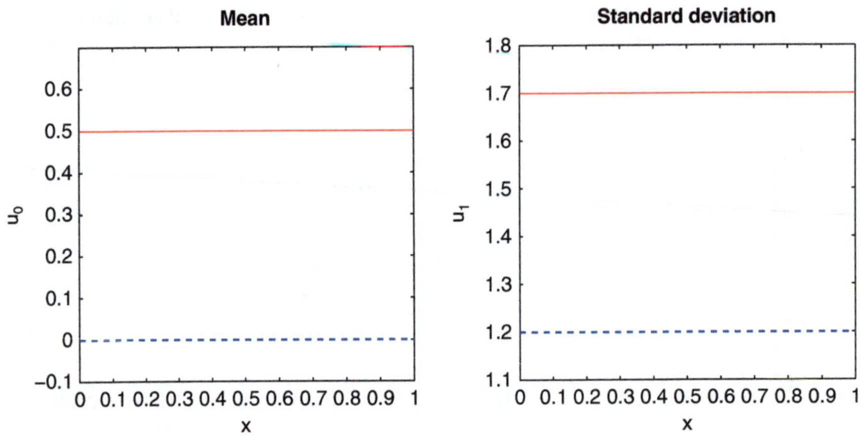

Fig. 7.12 u_0 is held fixed. Numerical solution (*solid red*) and exact solution (*dashed blue*), $t = 5$

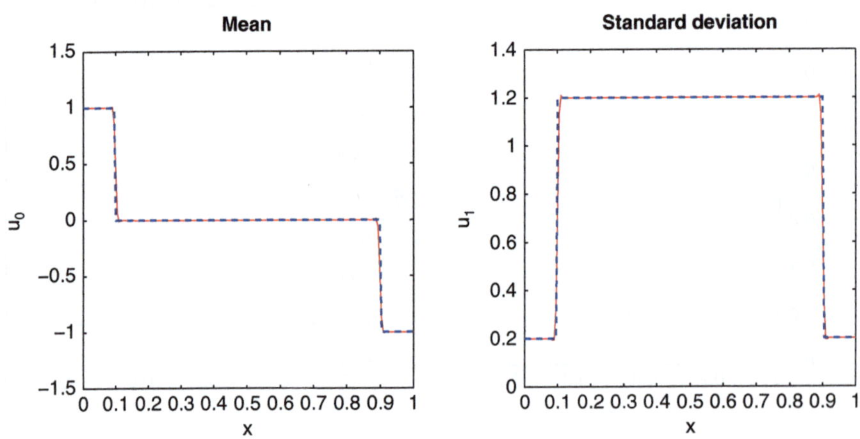

Fig. 7.13 u_0 extrapolated from the interior. Numerical solution (*solid red*) and exact solution (*dashed blue*), $t = 2$

7.1.3 Discussion of the Results with Incomplete Set of Data

The results in the preceding section are interesting and surprising. First, excellent results at steady-state (for long time) are obtained using the extrapolation technique, probably due to the fact that only one boundary condition is needed at the left boundary for $t > 2.5$.

By guessing data of the mean value and the variance, poor results are obtained. The impact of the error in the variance term (u_1) suggests that in a stochastic Galerkin system of order $M > 1$, the higher-order modes may be very important.

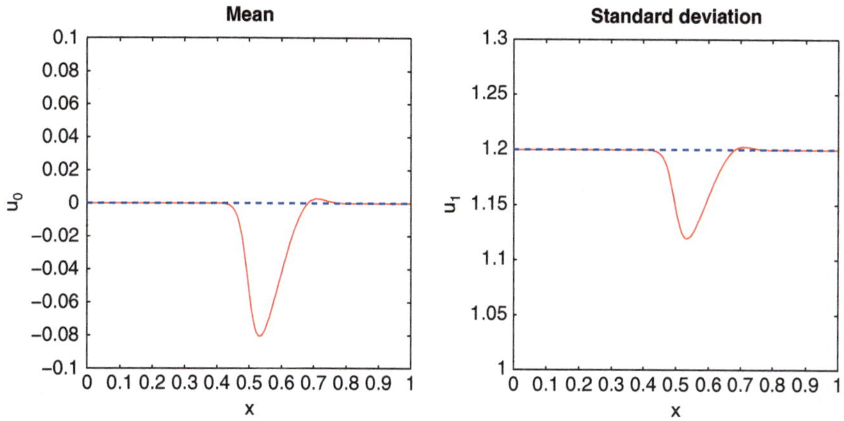

Fig. 7.14 u_0 extrapolated from the interior. Numerical solution (*solid red*) and exact solution (*dashed blue*), $t = 3$

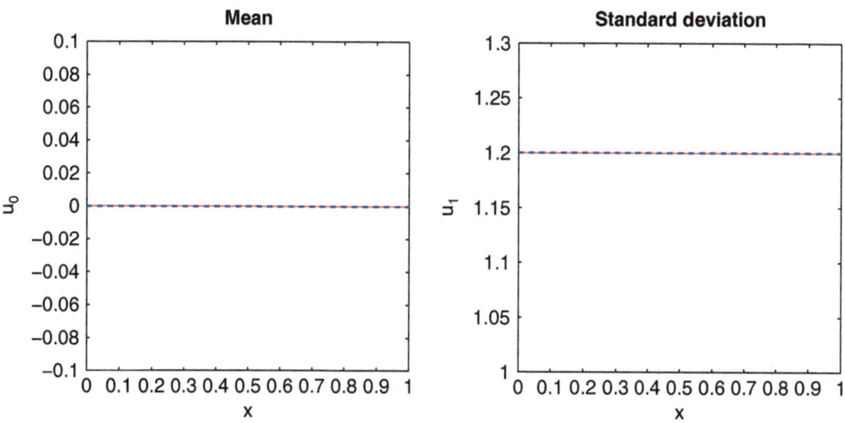

Fig. 7.15 u_0 extrapolated from the interior. Numerical solution (*solid red*) and exact solution (*dashed blue*), $t = 5$

The order of the error obtained here indicates that appropriate approximation of the higher-order terms is as important as guessing the expectation to get accurate results.

In many problems, sufficient data are not available to specify the correct number of variables. Unknown boundary values can then be constructed by extrapolation from the interior or by simply guessing the boundary data. We have investigated these two possible cases and for this specific problem, the extrapolation technique was superior.

It was also found that missing data for the expectation were not more serious than the lack of data for the higher mode (approximating the variance). This finding casts new light on the data requirement for higher-order expansions.

7.2 Summary and Conclusions

Uncertainty in data on inflow boundaries will propagate into the domain of interest and affect the solution. We have analyzed the stochastic Burgers' equation with a focus on the availability of data for the boundary conditions. To facilitate understanding, we deal only with the truncated representation $u(x,t,\xi) = u_0\psi_0 + u_1\psi_1$. This means that all the stochastic variation is accounted for by the single gPC coefficient u1, and the standard deviation of the solution is simply $|u_1|$. With this simplified setup, we obtain a few combinations of general situations for the boundary data: known expectation but unknown standard deviation, unknown expectation and standard deviation, etc. In the cases where we did not have available data, we remedy the situation by (i) guessing the data (expectation and/or standard deviation) or (ii) using extrapolation. The implications in all these situations on well-posedness, stability and accuracy are discussed.

In a general hyperbolic problem, the imposition of correct time-dependent boundary conditions will probably prove to be one of the more significant problems with the stochastic Galerkin method. A detailed investigation is necessary to find ways around the lack of time-dependent stochastic boundary data, especially for the higher moments. Most likely, special non-reflecting boundary conditions must be developed.

7.3 Exercises

Problem 7.1. Assume that you are given statistics in terms of mean value and standard deviation for the stochastic Galerkin Burgers' equation. You have reason to believe that higher-order coefficients are non-negligible. For a given value of standard deviation at the boundaries, what is the effect over time on the standard deviation over the interior domain? Use the supplied Matlab scripts for Burgers' equation, use second-order polynomial chaos (three terms), and try $u_1 = b, u_2 = 0$. Then try $u_1 = 0, u_2 = b$. Note that the standard deviation at the boundaries is identical for the two cases. Use the same initial values for u_0 for both cases. What do you observe for the interior domain standard deviation?

Problem 7.2. Analyze the problem illustrated in Sect. 7.1 by varying b (the standard deviation). Consider the case of $b = 0.05$ and $b = 0.5$.

Reference

1. Pettersson P, Iaccarino G, Nordström J (2010) Boundary procedures for the time-dependent Burgers' equation under uncertainty. Acta Math Sci 30(2):539–550. doi:10.1016/S0252-9602(10)60061-6, http://www.sciencedirect.com/science/article/pii/S0252960210600616

Part III
Euler Equations and Two-Phase Flow

Chapter 8
gPC for the Euler Equations

In many nonlinear applications of the stochastic Galerkin method, truncation of the generalized chaos expansion leads to non-unique formulations of the systems of equations. For instance, cubic products between stochastic quantities a, b and c are represented as products of truncated approximations $\tilde{a}$, $\tilde{b}$ and $\tilde{c}$, but the pseudospectral multiplication operator $*$ (to be explicitly defined in a later section), is not associative, i.e., $(\tilde{a} * \tilde{b}) * \tilde{c} \neq \tilde{a} * (\tilde{b} * \tilde{c})$. Similar problems are investigated in more detail in [2]. It is common practice to introduce these pseudospectral approximations since they imply a reduced numerical cost. Examples in the context of polynomial chaos for fluid flow include [13, 14].

The need to introduce stochastic expansions of inverse quantities, or square-roots of stochastic quantities of interest, adds to the number of different ways possible to approximate the original stochastic problem. This approximation leads to ambiguity of the problem formulation. We present a method where this ambiguity is avoided. Our formulation relies on a variable transformation where the square root of the density is computed, a computation that can be performed in a robust way in a small number of operations.

Poëtte et al. [7] used a nonlinear projection method to bound the oscillations close to stochastic discontinuities by PC expansion of the entropy variables obtained from a transformation of the conservative variables. Each time-step is complemented by a functional minimization to obtain the entropy variables needed to update the solution vector. The method we will present here may appear similar at first sight, but it relies on a different kind of variable transformation and not on kinetic theory considerations. We do not suggest a variable transformation for general conservation laws, but rather a formulation that specifically targets the solution of the Euler equations with uncertainty in the variables. It is less complicated than a direct gPC expansion of the conservative variables.

In the method presented, the system of equations is reformulated using Roe variables so that only quadratic terms occur. This is attractive since no fourth-order

M.P. Pettersson et al., *Polynomial Chaos Methods for Hyperbolic Partial Differential Equations*, Mathematical Engineering,
DOI 10.1007/978-3-319-10714-1_8

tensors need be approximated or calculated, resulting in increased accuracy and reduced computational cost. Moreover, there is no need to compute additional generalized chaos expansions for inverse quantities. The Roe variable expansion provides a simple and unambiguous formulation of the Euler equations. For brevity of notation, we will refer to this expansion method as the Roe expansion, and the method based on expansion of the conservative variables as the conservative expansion.

We consider the Sod test case subject to uncertainty in the density, and uncertain diaphragm location, respectively. The uncertainty is represented with a multiwavelet (MW) expansion in the stochastic dimension, following the framework outlined earlier in [3]. Multiwavelets are suitable for this problem since we need to represent discontinuities (localized support of basis functions) and still want high-order resolution in regions away from the discontinuities. Special cases of the MW basis include the Legendre polynomials and the piecewise constant Haar wavelets. The stochastic Galerkin system is obtained by projection of the stochastic Euler equations onto the MW basis functions.

Stochastic hyperbolic problems in general require a large number of stochastic basis functions for accurate representation. In particular, this problem becomes severe at large times [16]. One remedy is to use an adaptive stochastic basis that evolves in space and time to save computational cost. In the context of stochastic Galerkin methods for hyperbolic problems, Tryoen et al. introduced an adaptive method where the resolution was determined locally based on numerical estimates of the smoothness of the solution [12]. We will restrict ourselves to a non-adaptive stochastic basis and focus on the numerical solver rather than on the stochastic representation. This chapter is based on the work in [6].

8.1 Euler Equations with Input Uncertainty

Consider the 1D Euler equations, in non-dimensional form given by

$$\boldsymbol{u}_t + \boldsymbol{f}(\boldsymbol{u})_x = 0, \quad 0 \le x \le 1, t > 0, \tag{8.1}$$

where the solution and flux vector are given by

$$\boldsymbol{u} = \begin{bmatrix} \rho \\ \rho v \\ E \end{bmatrix}, \quad \boldsymbol{f} = \begin{bmatrix} \rho v \\ \rho v^2 + p \\ (E + p)v \end{bmatrix},$$

where ρ is density, v velocity, E total energy, and p pressure. A perfect gas equation of state is assumed, and energy and pressure are related by

$$E = \frac{p}{\gamma - 1} + \frac{1}{2}\rho v^2,$$

where γ is the ratio of the specific heats. For the numerical method, we need the flux Jacobian, given by

$$\frac{\partial \boldsymbol{f}}{\partial \boldsymbol{u}} = \begin{bmatrix} 0 & 1 & 0 \\ \frac{1}{2}(\gamma - 3)v^2 & (3-\gamma)v & \gamma - 1 \\ \frac{1}{2}(\gamma - 1)v^3 - vH & H - (\gamma - 1)v^2 & \gamma v \end{bmatrix},$$

with the total enthalpy $H = (E + p)/\rho$.

We scale the physical variables to get the dimensionless variables $\rho = \rho'/\rho'_{ref}$, $E = E'/(\gamma p'_{ref})$, $p = p'/(\gamma p'_{ref})$ and $v = v'/a'_{ref}$, where $a' = (\gamma p'/\rho')^{1/2}$ and the subscript ref denotes a reference state.

8.1.1 Formulation in Roe Variables

For the purpose of the design of an efficient numerical method, Roe [10] introduced the variables

$$\boldsymbol{w} = \begin{bmatrix} w_1 \\ w_2 \\ w_3 \end{bmatrix} = \begin{bmatrix} \rho^{1/2} \\ \rho^{1/2} v \\ \rho^{1/2} H \end{bmatrix}.$$

The flux and the conservative variables are given by

$$\hat{\boldsymbol{f}}(\boldsymbol{w}) = \begin{bmatrix} w_1 w_2 \\ \frac{\gamma-1}{\gamma} w_1 w_3 + \frac{\gamma+1}{2\gamma} w_2^2 \\ w_2 w_3 \end{bmatrix}, \quad \boldsymbol{u} = \hat{\boldsymbol{g}}(\boldsymbol{w}) = \begin{bmatrix} w_1^2 \\ w_1 w_2 \\ \frac{w_1 w_3}{\gamma} + \frac{\gamma-1}{2\gamma} w_2^2 \end{bmatrix}.$$

Then, the Euler equations in Roe's variables,

$$\hat{\boldsymbol{g}}(\boldsymbol{w})_t + \hat{\boldsymbol{f}}_x(\boldsymbol{w}) = 0 \tag{8.2}$$

is equivalent to (8.1). The flux Jacobian in the Roe variables is given by

$$\frac{\partial \hat{\boldsymbol{f}}}{\partial \boldsymbol{w}} = \begin{bmatrix} w_2 & w_1 & 0 \\ \frac{\gamma-1}{\gamma} w_3 & \frac{\gamma+1}{\gamma} w_2 & \frac{\gamma-1}{\gamma} w_1 \\ 0 & w_3 & w_2 \end{bmatrix}.$$

8.1.2 Stochastic Galerkin Formulation of the Euler Equations

In order to simplify the notation henceforth, we let the index of the MW series expansion start from 1. Define the pseudospectral product $u * v$ of order $M = M(N_p, N_r)$ by

$$(u * v)_k = \sum_{i=1}^{M} \sum_{j=1}^{M} u_i v_j \langle \psi_i \psi_j \psi_k \rangle, \qquad k = 1, \dots, M,$$

where

$$\langle \psi_i \psi_j \psi_k \rangle = \int_{\Omega} \psi_i(\xi) \psi_j(\xi) \psi_k(\xi) d\mathscr{P}.$$

Alternatively, using matrix notation as in the previous chapters, we can write the vector of coefficients of the spectral product $u * v$ as $\boldsymbol{A}(\boldsymbol{u})\boldsymbol{v}$, where

$$[\boldsymbol{A}(\boldsymbol{u})]_{jk} = \sum_{i=1}^{M} u_i \langle \psi_i \psi_j \psi_k \rangle. \tag{8.3}$$

We will need the pseudospectral inverse q^{-*}, defined as the solution of $q * q^{-*} = 1$, and the pseudospectral square root, defined as the solution $q^{*/2}$ of $q^{*/2} * q^{*/2} = q$, where the spectral expansion of the quantity of interest q is assumed to be known. For more details, see [1].

Let $\boldsymbol{u}^M$ denote the vector of coefficients of the MW expansion of u of order $M = M(N_p, N_r)$. M may take the same value for two distinct pairs of (N_p, N_r), but this ambiguity in notation will not matter in the derivation of the numerical method, for brevity we use only M in the superscripts. The Euler equations represented by the conservative formulation (8.1) can be written as an augmented system, after stochastic Galerkin projection,

$$\boldsymbol{u}_t^M + \boldsymbol{f}^M(\boldsymbol{u}^M)_x = 0, \tag{8.4}$$

where

$$\boldsymbol{u}^M = \begin{bmatrix} \boldsymbol{u}_1^M \\ \boldsymbol{u}_2^M \\ \boldsymbol{u}_3^M \end{bmatrix} = \begin{bmatrix} [(u_1)_1, \dots, (u_1)_M]^T \\ [(u_2)_1, \dots, (u_2)_M]^T \\ [(u_3)_1, \dots, (u_3)_M]^T \end{bmatrix},$$

$$\boldsymbol{f}^M(\boldsymbol{u}^M) = \begin{bmatrix} \boldsymbol{u}_2^M \\ (\boldsymbol{u}_1^M)^{-*} * \boldsymbol{u}_2^M * \boldsymbol{u}_2^M + p^M \\ (\boldsymbol{u}_3^M + p^M) * \boldsymbol{u}_2^M * (\boldsymbol{u}_1^M)^{-*} \end{bmatrix},$$

with $\boldsymbol{p}^M = (\gamma - 1)(\boldsymbol{u}_3^M - (\boldsymbol{u}_1^M)^{-*} * \boldsymbol{u}_2^M * \boldsymbol{u}_2^M/2)$. The cubic products of (8.4) are approximated by the application of two third-order tensors instead of one fourth-order tensor. That is, we replace $(a * b * c)_l = \sum_{ijk} \langle \psi_i \psi_j \psi_k \psi_l \rangle a_i b_j c_k$ by the approximation $(a * b * c)_l \approx ((a * b) * c)_l$. This approximation introduces an error in addition to the error from truncation of the gPC series to a finite number of terms. The effect of the error introduced by the approximation of higher-order tensors with successive application of third-order tensors was studied in [1], where it was found that the error is negligible if sufficiently high-order gPC expansions are used. We use this approximation with the conservative variables to make a fair comparison of the computational cost with the method we propose based on Roe variables.

For the Roe variable formulation, the stochastic Galerkin projection of (8.2) gives the system

$$\hat{\boldsymbol{g}}^M(\boldsymbol{w}^M)_t + \hat{\boldsymbol{f}}^M(\boldsymbol{w}^M)_x = 0, \tag{8.5}$$

where

$$\hat{\boldsymbol{g}}^M(\boldsymbol{w}^M) = \begin{bmatrix} \boldsymbol{w}_1^M * \boldsymbol{w}_1^M \\ \boldsymbol{w}_1^M * \boldsymbol{w}_2^M \\ \frac{\boldsymbol{w}_1^M * \boldsymbol{w}_3^M}{\gamma} + \frac{\gamma-1}{2\gamma} \boldsymbol{w}_2^M * \boldsymbol{w}_2^M \end{bmatrix},$$

$$\hat{\boldsymbol{f}}^M(\boldsymbol{w}^M) = \begin{bmatrix} \boldsymbol{w}_1^M * \boldsymbol{w}_2^M \\ \frac{\gamma-1}{\gamma} \boldsymbol{w}_1^M * \boldsymbol{w}_3^M + \frac{\gamma+1}{2\gamma} \boldsymbol{w}_2^M * \boldsymbol{w}_2^M \\ \boldsymbol{w}_2^M * \boldsymbol{w}_3^M \end{bmatrix}.$$

The flux Jacobian for the stochastic Galerkin system in the Roe variables is given by

$$\frac{\partial \hat{\boldsymbol{f}}^M}{\partial \boldsymbol{w}^M} = \begin{bmatrix} \boldsymbol{A}(\boldsymbol{w}_2^M) & \boldsymbol{A}(\boldsymbol{w}_1^M) & \boldsymbol{0}_{M\times M} \\ \frac{\gamma-1}{\gamma}\boldsymbol{A}(\boldsymbol{w}_3^M) & \frac{\gamma+1}{\gamma}\boldsymbol{A}(\boldsymbol{w}_2^M) & \frac{\gamma-1}{\gamma}\boldsymbol{A}(\boldsymbol{w}_1^M) \\ \boldsymbol{0}_{M\times M} & \boldsymbol{A}(\boldsymbol{w}_3^M) & \boldsymbol{A}(\boldsymbol{w}_2^M) \end{bmatrix}. \tag{8.6}$$

As $M \to \infty$, the formulations (8.4) and (8.5), as well as any other consistent formulation, are equivalent. However, M is assumed to be small (<20), and truncation and conditioning of the system matrices will play an important role in the accuracy of the solution.

We assume that γ is a deterministic constant in the formulation of the numerical schemes. Although it would imply additional pseudospectral multiplications, accounting for a random γ is a straightforward extension of the presented framework. This amounts to forming $\boldsymbol{A}(\gamma^{-1})$, which can be precomputed and stored for use in the updates of the numerical fluxes.

8.2 Numerical Method

We use MUSCL (Monotone Upstream-centered Schemes for Conservation Laws), introduced in [15]. For clarity of comparison of the numerical results, the MUSCL scheme is used for both the conservative variable formulation and the Roe variable formulation.

8.2.1 Expansion of Conservative Variables

Let m be the number of spatial cells and the uniform step length $\Delta x = 1/m$ and let $\vec{\boldsymbol{u}}^M$ be the spatial discretization of $\boldsymbol{u}^M$. The semidiscretized form of (8.4) is given by

$$\frac{d\boldsymbol{u}_j^M}{dt} + \frac{\boldsymbol{F}_{j+1/2}^M - \boldsymbol{F}_{j-1/2}^M}{\Delta x} = 0, \quad j = 1, \ldots, m, \tag{8.7}$$

where $\boldsymbol{F}_{j+1/2}^M$ denotes the numerical flux function evaluated at the interface between cells j and $j+1$.

For the MUSCL scheme with slope limited states $\vec{\boldsymbol{u}}^L$ and $\vec{\boldsymbol{u}}^R$, we take the numerical flux

$$\boldsymbol{F}_{j+\frac{1}{2}}^M = \frac{1}{2}\left(\boldsymbol{f}^M(\boldsymbol{u}_{j+\frac{1}{2}}^L) + \boldsymbol{f}^M(\boldsymbol{u}_{j+\frac{1}{2}}^R)\right) + \frac{1}{2}|(\tilde{\boldsymbol{J}}_c^M)_{j+\frac{1}{2}}|\left(\boldsymbol{u}_{j+\frac{1}{2}}^L - \boldsymbol{u}_{j+\frac{1}{2}}^R\right), \tag{8.8}$$

where the Roe average $\tilde{J}_c^M$ is the pseudospectral generalization of the standard Roe average of the deterministic Euler equations, i.e.,

$$\tilde{J}_c^M(\overline{\boldsymbol{v}}, \overline{\boldsymbol{H}}) = \begin{bmatrix} \boldsymbol{0}_{M\times M} & \boldsymbol{I}_{M\times M} & \boldsymbol{0}_{M\times M} \\ \frac{1}{2}(\gamma-3)\boldsymbol{A}(\overline{\boldsymbol{v}})^2 & (3-\gamma)\boldsymbol{A}(\overline{\boldsymbol{v}}) & (\gamma-1)\boldsymbol{I}_{M\times M} \\ \frac{1}{2}(\gamma-1)\boldsymbol{A}(\overline{\boldsymbol{v}})^3 - \boldsymbol{A}(\overline{\boldsymbol{v}})\boldsymbol{A}(\overline{\boldsymbol{H}}) & \boldsymbol{A}(\overline{\boldsymbol{H}}) - (\gamma-1)\boldsymbol{A}(\overline{\boldsymbol{v}})^2 & \gamma\boldsymbol{A}(\overline{\boldsymbol{v}}) \end{bmatrix},$$

where

$$\overline{\boldsymbol{v}} = (\rho_L^{-*/2} + \rho_R^{-*/2}) * (\rho_L^{*/2} * v_L + \rho_R^{*/2} * v_R),$$

and

$$\overline{\boldsymbol{H}} = (\rho_L^{*/2} * H_L + \rho_R^{*/2} * H_R) * (\rho_L^{-*/2} + \rho_R^{-*/2}).$$

The computation of $\overline{\boldsymbol{v}}$ and $\overline{\boldsymbol{H}}$ requires the spectral square root $\rho^{*/2}$ and its inverse, that are computed solving a nonlinear and a linear system, respectively.

Further details about the formulation of the Roe average matrix are given in [13]. The scheme is a direct generalization of the deterministic MUSCL scheme. Flux limiters are applied componentwise to all MW coefficients in sharp regions. For a more detailed description of the MUSCL scheme and flux limiters, see e.g. [4], and for application to the stochastic Burgers' equation, see [5].

8.2.2 Expansion of Roe's Variables

Let $\vec{w}^M = ((w_1^M)^T, (w_2^M)^T, \ldots, (w_m^M)^T)^T$ denote the spatial discretization of w^M. The semidiscretized form of (8.5) is given by

$$\frac{\partial \hat{g}^M(\vec{w}_j^M)}{\partial t} + \frac{\hat{F}^M_{j+1/2} - \hat{F}^M_{j-1/2}}{\Delta x} = 0, \quad j = 1, \ldots, m,$$

with the numerical flux function

$$\hat{F}_{j+\frac{1}{2}} = \frac{1}{2}\left(\hat{f}^M(w^L_{j+\frac{1}{2}}) + \hat{f}^M(w^R_{j+\frac{1}{2}})\right) + \frac{1}{2}|\tilde{J}^M_{j+\frac{1}{2}}|\left(w^L_{j+\frac{1}{2}} - w^R_{j+\frac{1}{2}}\right), \tag{8.9}$$

where $\tilde{J}^M = \tilde{J}^M(w^M)$ is the Roe matrix for the stochastic Galerkin formulation of the Euler equations in Roe's variables, to be derived below.

Each time-step provides the update of the solution vector $\hat{g}_j^M = \hat{g}^M(w_j^M)$, $j = 1, \ldots, m$, from which we can solve for $\vec{w}^M$ to be used in the update of the numerical flux. This involves solving the nonlinear systems

$$A(w^M_{1,j})w^M_{1,j} = \hat{g}^M_{1,j}, \quad j = 1, \ldots, m \tag{8.10}$$

for $w^M_{1,j}$, and then using $w^M_{1,j}$ to solve the linear $M \times M$-systems

$$A(w^M_{1,j})w^M_{2,j} = \hat{g}^M_{2,j}, \quad j = 1, \ldots, m$$

for $w^M_{2,j}$, and

$$A(w^M_{1,j})W^M_{3,j} = \gamma \hat{g}^M_{3,j} - \frac{\gamma - 1}{2} A(w^M_{2,j})w^M_{2,j}, \quad j = 1, \ldots, m$$

for $w^M_{3,j}$.

The system (8.10) is solved iteratively with a trust-region-dogleg algorithm.[1] Starting with the value of the previous time-step as the initial guess, few iterations are required (typically 2–3). The same method is used to solve for spectral square roots in the conservative variable formulation.

[1]This is the default algorithm for `fsolve` in Matlab. For more details, see [8].

8.2.3 Stochastic Galerkin Roe Average Matrix for Roe Variables

The Roe average matrix $\tilde{\boldsymbol{J}}^M$ is given as a function of the Roe variables $\boldsymbol{w}^M = ((\boldsymbol{w}_1^M)^T\ (\boldsymbol{w}_2^M)^T\ (\boldsymbol{w}_3^M)^T)^T$, where each $\boldsymbol{w}_i^M$ $(i = 1,2,3)$ is a vector of generalized chaos coefficients. It is designed to satisfy the following properties:

(i) $\tilde{\boldsymbol{J}}^M(\boldsymbol{w}^L, \boldsymbol{w}^R) \to \left.\frac{\partial \hat{\boldsymbol{f}}^M}{\partial w}\right|_{w=w'}$ as $\boldsymbol{w}^L, \boldsymbol{w}^R \to \boldsymbol{w}'$.

(ii) $\tilde{\boldsymbol{J}}^M(\boldsymbol{w}^L, \boldsymbol{w}^R) \times (\boldsymbol{w}^L - \boldsymbol{w}^R) = \hat{\boldsymbol{f}}^M(\boldsymbol{w}^L) - \hat{\boldsymbol{f}}^M(\boldsymbol{w}^R)$, $\forall \boldsymbol{w}^L, \boldsymbol{w}^R$.

(iii) $\tilde{\boldsymbol{J}}^M$ is diagonalizable with real eigenvalues and linearly independent eigenvectors.

In the standard approach introduced by Roe and commonly used for deterministic calculations, the conservative variables are mapped to the $\boldsymbol{w}$ variables, which are then averaged.

In the *deterministic* case, we have

$$\hat{\boldsymbol{f}}^L - \hat{\boldsymbol{f}}^R = \tilde{\boldsymbol{J}}(\boldsymbol{w}^L, \boldsymbol{w}^R) \times (\boldsymbol{w}^L - \boldsymbol{w}^R), \tag{8.11}$$

where

$$\tilde{\boldsymbol{J}}(\boldsymbol{w}^L, \boldsymbol{w}^R) = \begin{bmatrix} \overline{w}_2 & \overline{w}_1 & 0 \\ \frac{\gamma-1}{\gamma}\overline{w}_3 & \frac{\gamma+1}{\gamma}\overline{w}_2 & \frac{\gamma-1}{\gamma}\overline{w}_1 \\ 0 & \overline{w}_3 & \overline{w}_2 \end{bmatrix}.$$

Overbars denote arithmetic averages of assumed left and right values of a variable, i.e.,

$$\overline{w}_j = \frac{w_j^L + w_j^R}{2}, \quad j = 1,2,3.$$

It is a straightforward extension of the analysis by Roe in [10] to show properties (i) and (ii) for the Roe variables, without mapping to the conservative variables. To prove (iii) we note that there exists an eigenvalue decomposition

$$\tilde{\boldsymbol{J}} = \boldsymbol{V}\boldsymbol{D}\boldsymbol{V}^{-1}, \tag{8.12}$$

where

$$\boldsymbol{V} = \begin{bmatrix} \frac{w_1}{w_3} & \frac{w_1}{w_3} & -\frac{w_1}{w_3} \\ \frac{w_2 - \sqrt{w_2^2 + 8w_1w_3\gamma(\gamma-1)}}{2\gamma w_3} & \frac{w_2 + \sqrt{w_2^2 + 8w_1w_3\gamma(\gamma-1)}}{2\gamma w_3} & 0 \\ 1 & 1 & 1 \end{bmatrix}, \tag{8.13}$$

$$
\boldsymbol{D} = \begin{bmatrix} \frac{w_2(1+2\gamma)-\sqrt{8w_1w_3\gamma(\gamma-1)+w_2^2}}{2\gamma} & 0 & 0 \\ 0 & \frac{w_2(1+2\gamma)+\sqrt{8w_1w_3\gamma(\gamma-1)+w_2^2}}{2\gamma} & 0 \\ 0 & 0 & w_2 \end{bmatrix}. \tag{8.14}
$$

The eigenvalues of $\tilde{\boldsymbol{J}}$ are real and distinct, so property (iii) is also satisfied.

Now consider the stochastic Galerkin formulation, i.e., assume that the w_i's are vectors of generalized chaos coefficients. The stochastic Galerkin Roe average matrix $\tilde{\boldsymbol{J}}^M$ for the Roe variables formulation is a generalization of the mapping (8.11), i.e., of the matrix $\tilde{\boldsymbol{J}}$. We define

$$
\tilde{\boldsymbol{J}}^M(\boldsymbol{w}^L, \boldsymbol{w}^R) = \tilde{\boldsymbol{J}}^M(\overline{\boldsymbol{w}}) = \begin{bmatrix} \boldsymbol{A}(\overline{\boldsymbol{w}}_2) & \boldsymbol{A}(\overline{\boldsymbol{w}}_1) & \boldsymbol{0}_{M\times M} \\ \frac{\gamma-1}{\gamma}\boldsymbol{A}(\overline{\boldsymbol{w}}_3) & \frac{\gamma+1}{\gamma}\boldsymbol{A}(\overline{\boldsymbol{w}}_2) & \frac{\gamma-1}{\gamma}\boldsymbol{A}(\overline{\boldsymbol{w}}_1) \\ \boldsymbol{0}_{M\times M} & \boldsymbol{A}(\overline{\boldsymbol{w}}_3) & \boldsymbol{A}(\overline{\boldsymbol{w}}_2) \end{bmatrix}, \tag{8.15}
$$

where the submatrix $\boldsymbol{A}(\boldsymbol{w}_j)$ is given by (8.3) and $\overline{\boldsymbol{w}} = (\boldsymbol{w}^L + \boldsymbol{w}^R)/2$.

Proposition 8.1. *Property (i) is satisfied by (8.15).*

Proof. With $\boldsymbol{w}^L = \boldsymbol{w}^R = \boldsymbol{w}'$, $\tilde{\boldsymbol{J}}^M(\boldsymbol{w}^L, \boldsymbol{w}^R) = \tilde{\boldsymbol{J}}^M(\boldsymbol{w}', \boldsymbol{w}') = \left.\frac{\partial \hat{\boldsymbol{f}}^M}{\partial \boldsymbol{w}^M}\right|_{\boldsymbol{w}=\boldsymbol{w}'}$ by (8.6).

Proposition 8.2. *Property (ii) is satisfied by (8.15).*

Proof.

$$
\begin{aligned}
\tilde{\boldsymbol{J}}^M(\boldsymbol{w}^L, \boldsymbol{w}^R) \times (\boldsymbol{w}^L - \boldsymbol{w}^R) &= \frac{1}{2}\left(\tilde{\boldsymbol{J}}^M(\boldsymbol{w}^L) + \tilde{\boldsymbol{J}}^M(\boldsymbol{w}^R)\right)(\boldsymbol{w}^L - \boldsymbol{w}^R) \\
&= \frac{1}{2}\tilde{\boldsymbol{J}}^M(\boldsymbol{w}^L)\boldsymbol{w}^L - \frac{1}{2}\tilde{\boldsymbol{J}}^M(\boldsymbol{w}^R)\boldsymbol{w}^R = \hat{\boldsymbol{f}}^M(\boldsymbol{w}^L) - \hat{\boldsymbol{f}}^M(\boldsymbol{w}^R),
\end{aligned} \tag{8.16}
$$

where the last equality follows from the fact that the stochastic Galerkin generalizations of the Euler equations are homogeneous of degree 1.

To prove (iii), we will need the following proposition.

Lemma 8.1. *Let $\boldsymbol{A}(\boldsymbol{w}_j)$ $(j = 1, 2, 3)$ be defined by (8.3) and $\boldsymbol{A}(\boldsymbol{w}_j) = \boldsymbol{Q}\boldsymbol{\Lambda}_j\boldsymbol{Q}^T$ be an eigenvalue decomposition with constant eigenvector matrix $\boldsymbol{Q}$ and assume that $\boldsymbol{\Lambda}_1$ and $\boldsymbol{\Lambda}_3$ are non-singular. Then the stochastic Galerkin Roe average matrix $\tilde{\boldsymbol{J}}^M$ has an eigenvalue decomposition $\tilde{\boldsymbol{J}}^M = \boldsymbol{X}\tilde{\boldsymbol{\Lambda}}^M\boldsymbol{X}^{-1}$ with a complete set of eigenvectors.*

Proof. We will use the Kronecker product $\otimes$, defined for two matrices $\boldsymbol{B}$ (of size $m \times n$) and $\boldsymbol{C}$ by

$$
\boldsymbol{B} \otimes \boldsymbol{C} = \begin{bmatrix} b_{11}\boldsymbol{C} & \dots & b_{1n}\boldsymbol{C} \\ \vdots & \ddots & \vdots \\ b_{m1}\boldsymbol{C} & \dots & b_{mn}\boldsymbol{C} \end{bmatrix}.
$$

The eigenvalue decompositions of each $M \times M$ matrix block of (8.15) have the same eigenvector matrix $\boldsymbol{Q}$, hence we can write

$$\tilde{\boldsymbol{J}}^M = (\boldsymbol{I}_{3\times3} \otimes \boldsymbol{Q})\hat{\boldsymbol{J}}(I_{3\times3} \otimes \boldsymbol{Q}^T), \tag{8.17}$$

where

$$\hat{\boldsymbol{J}} = \begin{bmatrix} \boldsymbol{\Lambda}_2 & \boldsymbol{\Lambda}_1 & \boldsymbol{0}_{M\times M} \\ \frac{\gamma-1}{\gamma}\boldsymbol{\Lambda}_3 & \frac{\gamma+1}{\gamma}\boldsymbol{\Lambda}_2 & \frac{\gamma-1}{\gamma}\boldsymbol{\Lambda}_1 \\ \boldsymbol{0}_{M\times M} & \boldsymbol{\Lambda}_3 & \boldsymbol{\Lambda}_2 \end{bmatrix}.$$

By assumption, $\boldsymbol{I}_{3\times3} \otimes \boldsymbol{Q}$ is non-singular, and it remains to be shown that $\hat{\boldsymbol{J}}$ has distinct eigenvectors. Let

$$\boldsymbol{S} = \mathrm{diag}(\boldsymbol{\Lambda}_1\boldsymbol{\Lambda}_3^{-1}, \sqrt{(\gamma-1)/\gamma}\boldsymbol{\Lambda}_1^{1/2}\boldsymbol{\Lambda}_3^{-1/2}, \boldsymbol{I}_{M\times M}).$$

By assumption, $\boldsymbol{\Lambda}_1$ and $\boldsymbol{\Lambda}_3$ are invertible, so $\boldsymbol{S}$ and $\boldsymbol{S}^{-1}$ exist. We have

$$\boldsymbol{J}^S \equiv \boldsymbol{S}^{-1}\hat{\boldsymbol{J}}\boldsymbol{S} = \begin{bmatrix} \boldsymbol{\Lambda}_2 & \left[\frac{\gamma-1}{\gamma}\boldsymbol{\Lambda}_1\boldsymbol{\Lambda}_3\right]^{1/2} & \boldsymbol{0}_{M\times M} \\ \left[\frac{\gamma-1}{\gamma}\boldsymbol{\Lambda}_1\boldsymbol{\Lambda}_3\right]^{1/2} & \frac{\gamma-1}{\gamma}\boldsymbol{\Lambda}_2 & \left[\frac{\gamma-1}{\gamma}\boldsymbol{\Lambda}_1\boldsymbol{\Lambda}_3\right]^{1/2} \\ \boldsymbol{0}_{M\times M} & \left[\frac{\gamma-1}{\gamma}\boldsymbol{\Lambda}_1\boldsymbol{\Lambda}_3\right]^{1/2} & \boldsymbol{\Lambda}_2 \end{bmatrix}. \tag{8.18}$$

Clearly, $\boldsymbol{J}^S$ is symmetric and has the same eigenvalues as $\hat{\boldsymbol{J}}$ and $\tilde{\boldsymbol{J}}^M$. Hence, $\boldsymbol{J}^S$ has an eigenvalue decomposition $\boldsymbol{J}^S = \boldsymbol{Y}\tilde{\boldsymbol{\Lambda}}^M\boldsymbol{Y}^T$. Then,

$$\hat{\boldsymbol{J}} = \boldsymbol{S}\boldsymbol{Y}\tilde{\boldsymbol{\Lambda}}^M\boldsymbol{Y}^T\boldsymbol{S}^{-1} = \boldsymbol{S}\boldsymbol{Y}\tilde{\boldsymbol{\Lambda}}^M(\boldsymbol{S}\boldsymbol{Y})^{-1}. \tag{8.19}$$

Combining (8.17) and (8.19), we get

$$\tilde{\boldsymbol{J}}^M = [(\boldsymbol{I}_{3\times3} \otimes \boldsymbol{Q})\boldsymbol{S}\boldsymbol{Y}]\tilde{\boldsymbol{\Lambda}}^M[(\boldsymbol{I}_{3\times3} \otimes \boldsymbol{Q})\boldsymbol{S}\boldsymbol{Y}]^{-1}.$$

Setting $\boldsymbol{X} = (\boldsymbol{I}_{3\times3} \otimes \boldsymbol{Q})\boldsymbol{S}\boldsymbol{Y}$, we get the eigenvalue decomposition $\tilde{\boldsymbol{J}}^M = \boldsymbol{X}\tilde{\boldsymbol{\Lambda}}^M\boldsymbol{X}^{-1}$. By assumption, $\boldsymbol{S}$ and $\boldsymbol{Y}$ are non-singular, and we have

$$\det(\boldsymbol{X}) = \det((\boldsymbol{I}_{3\times3} \otimes \boldsymbol{Q})\boldsymbol{S}\boldsymbol{Y}) \neq 0,$$

which proves that $\boldsymbol{X}$ is non-singular, and thus $\tilde{\boldsymbol{J}}^M$ has a complete set of eigenvectors.

Proposition 8.3. *Property (iii) is satisfied by (8.15).*

Proof. Lemma 8.1 shows that since the eigenvalue matrix $\tilde{\boldsymbol{\Lambda}}^M$ is also the eigenvalue matrix of the symmetric matrix $\boldsymbol{J}^S$ defined in (8.18), the eigenvalues are all real. Lemma 8.1 also shows that the eigenvectors are distinct.

The conditions in Lemma 8.1 are true for certain basis functions assuming moderate stochastic variation, but the same can not be guaranteed for every case, and certainly does not hold for pathological cases with negative density, for example. The requirement of non-singularity of $\boldsymbol{\Lambda}_1, \boldsymbol{\Lambda}_3$ is not very restrictive since it amounts to excluding unphysical behavior, for instance naturally positive quantities taking negative values with nonzero probability. The assumption of constant eigenvectors of the matrix $\boldsymbol{A}$ holds for Haar wavelets (i.e., multiwavelets with $N_p = 0$), for all orders $M = 2^{N_r}$, with $N_r \in \mathbb{N}$. See Sect. B.1 for a proof sketch. Expressions for the first constant eigenvalue decompositions are included in Sect. B.2 for Haar wavelets and piecewise linear multiwavelets. The eigenvectors of $\boldsymbol{A}$ for $M = 1, 2, 4, 8$ are shown to be constant, but we do not give a proof that this is true for piecewise linear multiwavelets of any order M.

Remark 8.1. The Roe variable scheme has been outlined under the implicit assumption of uncertainty introduced in the initial and/or boundary conditions. However, situations such as uncertainty in the adiabatic coefficient γ may be treated in a similar way, although such treatment would result in additional pseudospectral products. Pseudospectral approximations of $(\gamma - 1)/\gamma$ and $(\gamma + 1)/\gamma$ could then be precomputed to sufficient accuracy.

Remark 8.2. For both the conservative variable formulation and the Roe variable formulation, we need to find the eigenvalue decomposition of $\tilde{\boldsymbol{J}}_c^M$ (or $\tilde{\boldsymbol{J}}^M$) at each time-step and each spatial point. For the cases of piecewise constant or piecewise linear MW, we can find this analytically and thus at low computational cost. For higher-order polynomial MW, we may rely on iterative methods for the eigenvalue decomposition of these $3M \times 3M$ subsystems. To this end, one may, for example, use the approximate low-order polynomial method that was introduced and successfully applied in [13] for very similar problems.

8.3 Numerical Results

We use the method of manufactured solutions to verify the second-order convergence in space of a smooth problem using the MUSCL scheme with Roe variables. We then introduce two test cases for the non-smooth problem; case 1 with an initial function that can be exactly represented by two Legendre polynomials, and case 2 with slow initial decay of the MW coefficients in both N_p and N_r. The errors in computed quantities of interest (here variances) as functions of the order of MW are investigated. Qualitative results are then presented to indicate the behavior we can expect for the convergence of two special cases of MW, namely the Legendre polynomials and Haar wavelet basis, respectively. Robustness with respect to more extreme cases (density close to zero leading to high Mach number) is demonstrated for the Roe variable formulation for a supersonic case where the conservative variable method breaks down. Finally, we perform a comparative study of the computational time for the formulation in conservative variables and the formulation in Roe variables.

8.3.1 Spatial Convergence

The MUSCL scheme with appropriate flux limiters is second-order accurate for smooth solutions. Since the Euler solution in general becomes discontinuous in finite time, the method of manufactured solutions [9, 11] is used to solve the Euler equations with source terms for a known smooth solution. The smooth solution is inserted into the Euler equations (8.1) and results in a nonzero right-hand side that is used as a source function. In order to test the capabilities of the method, we choose a solution that varies in space, time and in the stochastic dimension, and with time-dependent boundary conditions. It is designed to resemble a physical solution with non-negative density and pressure. The solution is given by

$$\begin{bmatrix} \rho \\ v \\ p \end{bmatrix} = \begin{bmatrix} \rho_0 + \rho_1 \tanh(s(x_0 - x + t + \sigma\xi)) \\ \tanh(s(x_0 + v_0 - x + t + \sigma\xi)) + \tanh(-s(x_0 - v_0 - x + t + \sigma\xi)) \\ p_0 + p_1 \tanh(s(x_0 - x + t + \sigma\xi)) \end{bmatrix}.$$

The parameters are set to $\rho_0 = p_0 = 0.75$, $\rho_1 = p_1 = x_0 = 0.25$, $v_0 = 0.05$, $s = 10, \sigma = 0.1$ and $\xi \in \mathcal{U}[-1, 1]$. The solution is shown in Fig. 8.1.

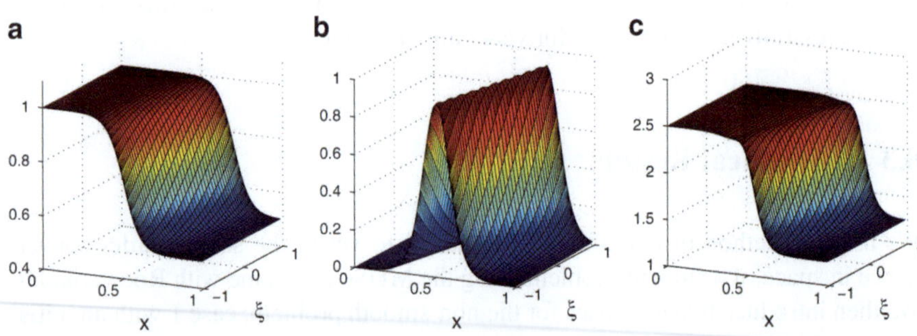

Fig. 8.1 Manufactured smooth solution as a function of x and ξ, $t = 0.15$. (**a**) Density. (**b**) Velocity. (**c**) Energy

We measure the error in the computed $\vec{u}(x,t,\xi)$ in the $L_2(\Omega,\mathscr{P})$ norm and the discrete ℓ_2 norm,

$$\begin{aligned}
\left\|\vec{u}^M - \vec{u}\right\|_{2,2} &\equiv \left\|\vec{u}^M - \vec{u}\right\|_{\ell_2,L_2(\Omega,\mathscr{P})} \\
&= \left(\Delta x \sum_{i=1}^{m} \left\|\boldsymbol{u}^M(x_i,t,\xi) - \boldsymbol{u}(x_i,t,\xi)\right\|^2_{L_2(\Omega,\mathscr{P})}\right)^{1/2} \\
&= \left(\Delta x \sum_{i=1}^{m} \int_{\Omega} (\boldsymbol{u}^M(x_i,t,\xi) - \boldsymbol{u}(x_i,t,\xi))^2 d\mathscr{P}(\xi)\right)^{1/2} \\
&\approx \left(\Delta x \sum_{i=1}^{m}\sum_{j=1}^{q} (\boldsymbol{u}^M(x_i,t,\xi_q^{(j)}) - \boldsymbol{u}(x_i,t,\xi_q^{(j)}))^2 w_q^{(j)}\right)^{1/2}, \quad (8.20)
\end{aligned}$$

where a q-point quadrature rule with points $\{\xi_q^{(j)}\}_{j=1}^q$ and weights $\{w_q^{(j)}\}_{j=1}^q$ was used in the last line to approximate the integral in ξ. The Gauss-Legendre quadrature is used here since the solution is smooth in the stochastic dimension.

Figure 8.2 depicts the spatial convergence in the $\|.\|_{2,2}$ norm of the error in density, velocity and energy. An order $(N_p, N_r) = (10, 0)$ basis is used to represent the uncertainty. The solution dynamics is initially concentrated in the left part of the spatial domain. By the time of $t = 0.4$, it has moved to the right and has begun to exit the spatial domain, so the time snapshots of Fig. 8.2 summarize the temporal history of the spatial error decay. The theoretical optimal convergence rate for the MUSCL scheme with the van Leer flux limiter is obtained for all times and all quantities.

8.3.2 Initial Conditions and Discontinuous Solutions

We consider (8.1) with two different initial functions on the domain [0, 1]. Since the analytical solution of Sod's test case is known for any fixed value of the input parameters, the exact stochastic solution can be formulated as a function of the stochastic input ξ. Exact statistics can be computed by numerical integration over ξ. As case number 1, assume that the density is subject to uncertainty, and all other quantities are deterministic at $t = 0$. The initial condition for (8.1) is given by

$$u(x,t=0,\xi) = \begin{cases} u_L = (1+\sigma\xi,\ 0,\ 2.5/\gamma)^T & x < 0.5 \\ u_R = (0.125(1+\sigma\xi),\ 0,\ 0.25/\gamma)^T & x > 0.5 \end{cases},$$

where we assume $\xi \in \mathscr{U}[-1,1]$, $\gamma = 1.4$ and the scaling parameter $\sigma = 0.5$. This is a simple initial condition in the sense that the first two Legendre polynomials are sufficient to represent the initial function exactly. As case number 2, we consider (8.1) subject to uncertainty in the initial shock location. Let

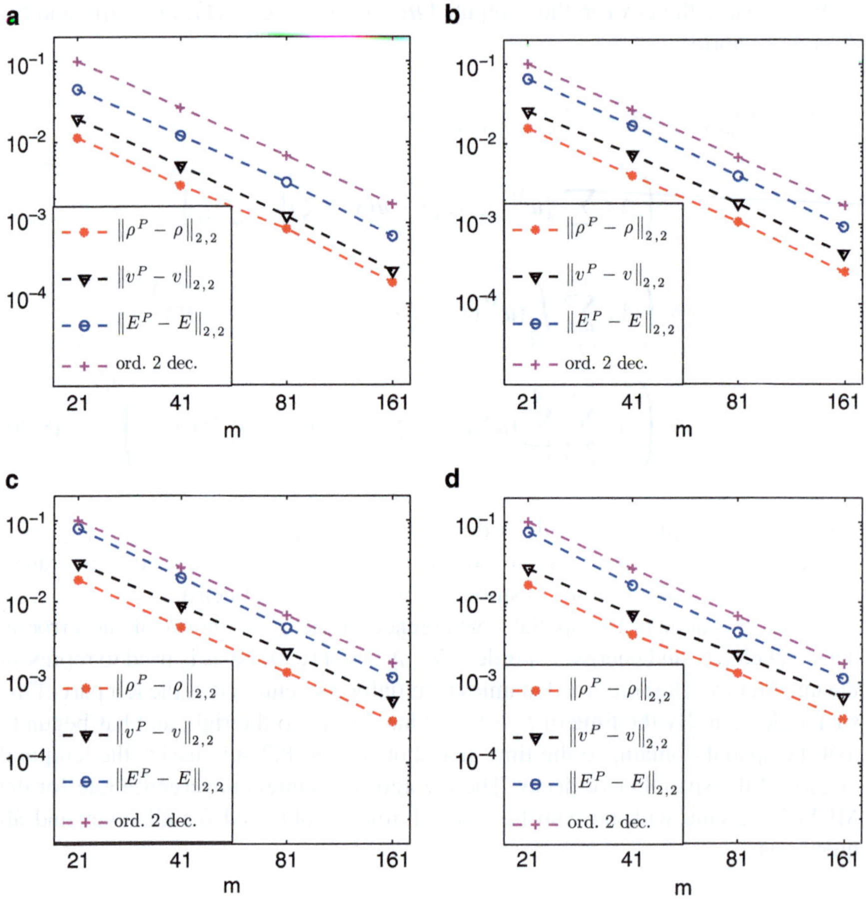

Fig. 8.2 Convergence in space using the method of manufactured solutions, $N_p = 10$, $N_r = 0$ (Legendre polynomials). Superscript P denotes the numerical pseudospectral solution. (**a**) $t = 0.05$. (**b**) $t = 0.1$. (**c**) $t = 0.2$. (**d**) $t = 0.4$

$$u(x, t = 0, \xi) = \begin{cases} u_L = (1,\ 0,\ 2.5/\gamma)^T & x < 0.5 + \sigma\eta \\ u_R = (0.125,\ 0,\ 0.25/\gamma)^T & x > 0.5 + \sigma\eta \end{cases},$$

where we assume $\gamma = 1.4$ and the scaling parameter $\sigma = 0.05$. Here, η takes a *triangular distribution*, which we parameterize as a nonlinear function in $\xi \in \mathscr{U}[-1, 1]$, i.e.,

$$\eta(\xi) = (-1 + \sqrt{\xi + 1})\mathbb{1}_{\{-1 \le \xi \le 0\}}(\xi) + (1 - \sqrt{1 - \xi})\mathbb{1}_{\{0 < \xi \le 1\}}(\xi),$$

where the indicator function $\mathbb{1}_{\{A\}}$ of a set A is defined by $\mathbb{1}_{\{A\}}(\xi) = 1$ if $\xi \in A$ and zero otherwise. For case 2, exact representation of the initial function requires

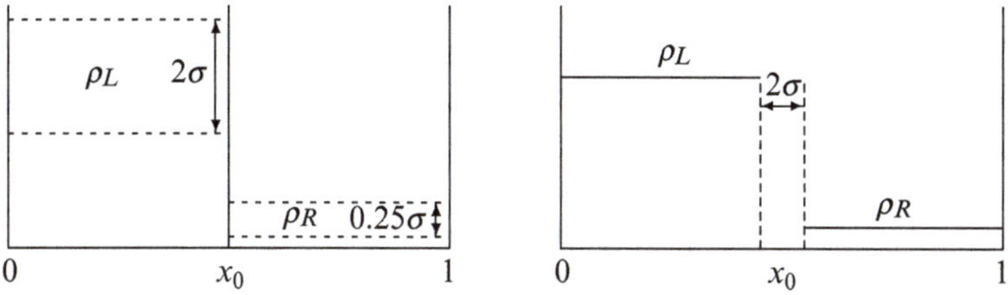

Fig. 8.3 Schematic representation of the initial setup for case 1 (*left*) and case 2 (*right*)

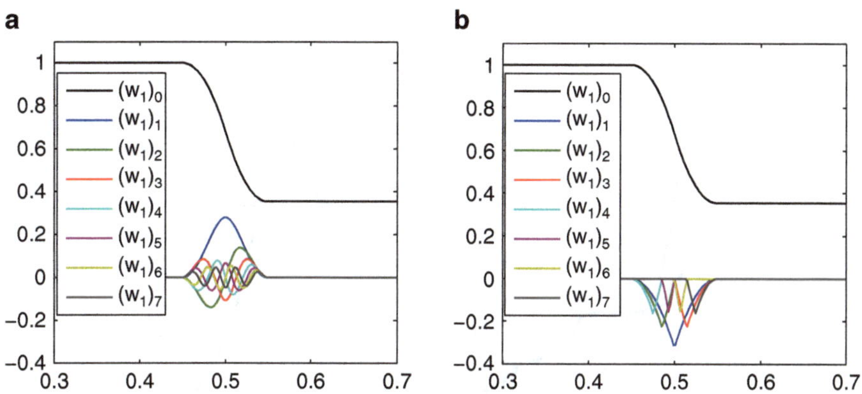

Fig. 8.4 Initial w_1 modes for case 2, first 8 basis functions. (**a**) Legendre polynomials. (**b**) Haar wavelets

an infinite number of expansion terms in the MW basis. Figure 8.3 depicts the shock tube setup for the two cases, with dashed lines denoting uncertain parameters. We will also investigate another version of case 2, where the right state density is significantly reduced to obtain a strong shock.

8.3.3 Spatial and Stochastic Resolution Requirements

For case 2, note that although the initial shock position can be exactly described by the first two terms of the Legendre polynomial chaos expansion, this is not the case for the initial state variables. In fact, for the polynomial chaos expansions of the density, momentum and energy, the error decays only slowly with the number of expansion terms. Thus, unless a reasonably large number of expansion terms is retained, the stochastic Galerkin solution of case 2 will not be accurate even for small times.

The Legendre coefficients at small times display an oscillating behavior that becomes sharper with the order of the coefficients. The wavelet coefficients exhibit peaks that get sharper with the resolution level, and require a fine mesh. Figure 8.4

shows the initial Legendre coefficients and the initial Haar wavelets for case 2. The numerical method has a tendency to smear the chaos coefficients, resulting in underprediction of the variance. The increasing cost of using a larger number of basis functions is further increased by the need for a finer mesh to resolve the solution modes.

Figure 8.5 shows the temporal evolution of the mean and variance of the density of case 2 as a function of space on a fine mesh of 500 spatial points and 32 piecewise linear multiwavelets. The mean and the variance are both reasonably well captured for this case. Figure 8.6 depicts case 2 for a similar setup, but with 32 Haar wavelets. The mean is well captured, but the variance is not fully captured. The

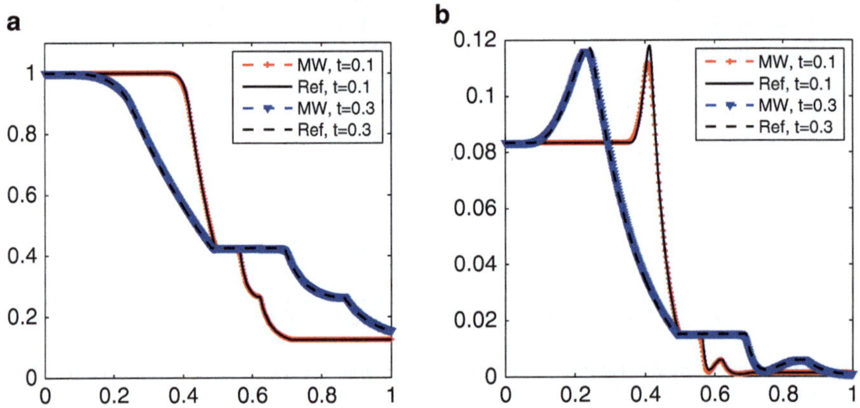

Fig. 8.5 Temporal evolution of the mean and variance of the density for case 1, using Roe variables, 500 spatial points and 32 piecewise linear multiwavelets. (**a**) Mean density. (**b**) Variance of density

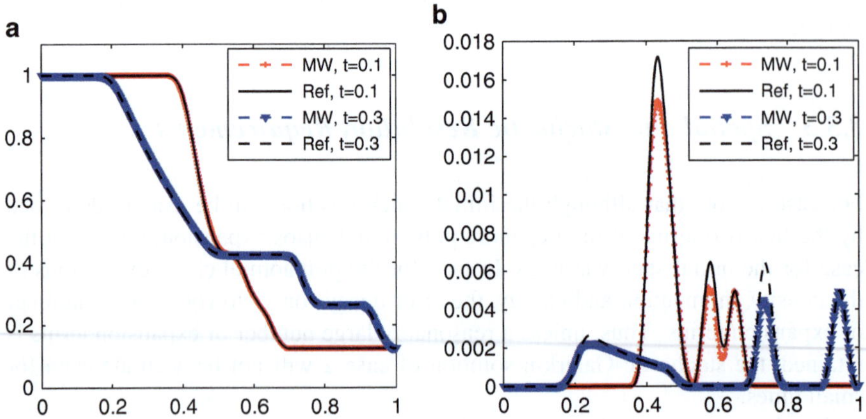

Fig. 8.6 Temporal evolution of the mean and variance of the density for case 2, using Roe variables, 500 spatial points and 32 Haar wavelets. (**a**) Mean density. (**b**) Variance of density

three variance peaks correspond to the rarefaction wave, contact discontinuity and the shock, respectively. As time progresses, the variance peaks will propagate out of the computational domain.

8.3.4 Convergence of Multiwavelet Expansions

For moderate simulation times, the numerical solution on a sufficiently fine spatial mesh converges as the order of MW expansion increases by increasing the polynomial degree N_p or the resolution level N_r. Figure 8.7 shows the decay in the

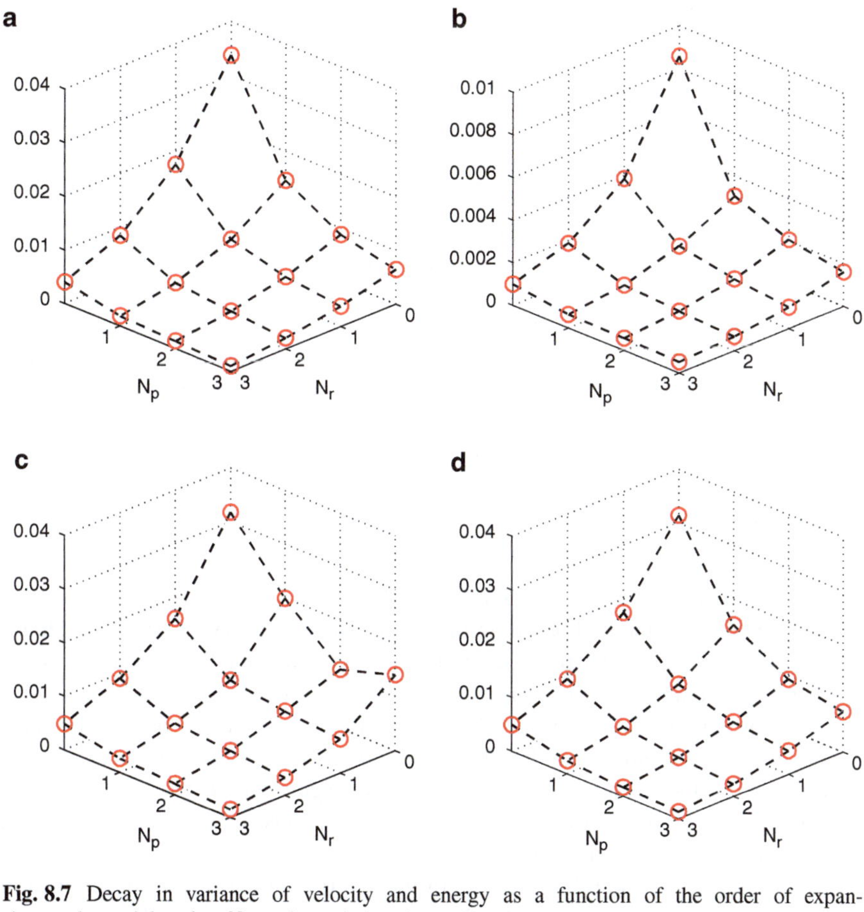

Fig. 8.7 Decay in variance of velocity and energy as a function of the order of expansion, polynomial order N_p and resolution level N_r. Case 1, $t = 0.05$, 280 spatial points restricted to $x \in [0.4, 0.65]$. Solution obtained with the Roe variable scheme. (**a**) Case 1, $\|Var(v^M) - Var(v)\|_2$. (**b**) Case 1, $\|Var(E^M) - Var(E)\|_2$. (**c**) Case 2, $\|Var(v^M) - Var(v)\|_2$. (**d**) Case 2, $\|Var(E^M) - Var(E)\|_2$

error of the variance of velocity and energy as a function of N_p and N_r. For well-behaved cases like these, one may freely choose between increasing N_p and N_r, in order to increase the accuracy of the solution of the quantity of interest.

For longer simulation times or more extreme cases, e.g., supersonic flow, high-order polynomial representation (increasing N_p) may not lead to increased accuracy, but rather to breakdown of the numerical method. Next, we study the qualitative properties of the MW representation of case 1 and case 2 for two extreme cases of MW parameters: Legendre polynomials ($N_r = 0$) and piecewise constant Haar wavelets ($N_p = 0$).

Figure 8.8 shows the density surface in the $x - \xi$-plane of case 1 and case 2 at $t = 0.15$ based on exact solution evaluations, and computed with Legendre polynomials and Haar wavelets. The computed solution with Legendre polynomial reconstruction captures essential features of the exact solution, but the use of global polynomials causes oscillations downstream of the shock.

With Haar wavelets, there are no oscillations downstream, unlike the Legendre polynomials case. However, the eight 'plateaus' seen in Fig. 8.8e correspond to the eight basis functions. When the order of wavelet chaos expansion increases, the number of plateaus increases, and the solution converges to the exact solution.

From Fig. 8.8, it is clear that the effect of the choice of multiwavelet basis depends to some extent on the problem at hand. Haar wavelets yield numerical solutions that are free of oscillations but converge only slowly. Oscillations around discontinuities in stochastic space should be expected when a polynomial basis is used and may lead to severe problems when variables attain unphysical values, e.g., when the oscillations downstream of the shock lead to negative density. Thus, more robust multiwavelets are required for problems with stronger shocks, as we demonstrate below.

8.3.5 *Robustness*

Complex supersonic test cases have already been successfully treated with a stochastic Galerkin method based on the conservative formulation, see for instance [14]. In general, the stochastic Galerkin method applied to the Roe variables gives a more robust method than the conservative variables formulation. The conservative formulation is more prone to ill-conditioning of the pseudospectral operations in the computation of the numerical flux. However, for cases where the matrix $\boldsymbol{A}$ in (8.3) has an eigenvalue decomposition with constant eigenvectors, the pseudospectral systems simplify to a series of scalar operations, thus avoiding ill-conditioned systems. An example is given in Sect. 8.3.6.

Figure 8.9 shows the relative errors of the solution in the 2, 2 norm (defined in (8.20)) for modified versions of case 2 with stronger shocks, obtained by increasing the difference between ρ_L and ρ_R. We fix $\rho_L = 1$, and let ρ_R take a range of different values, $\rho_R = 2^{-k}$, $k = 3, \ldots, 8$ for 8 basis wavelets. This corresponds to Mach numbers up to $Ma = 2.0$. Figure 8.9 also includes the relative error of the Mach

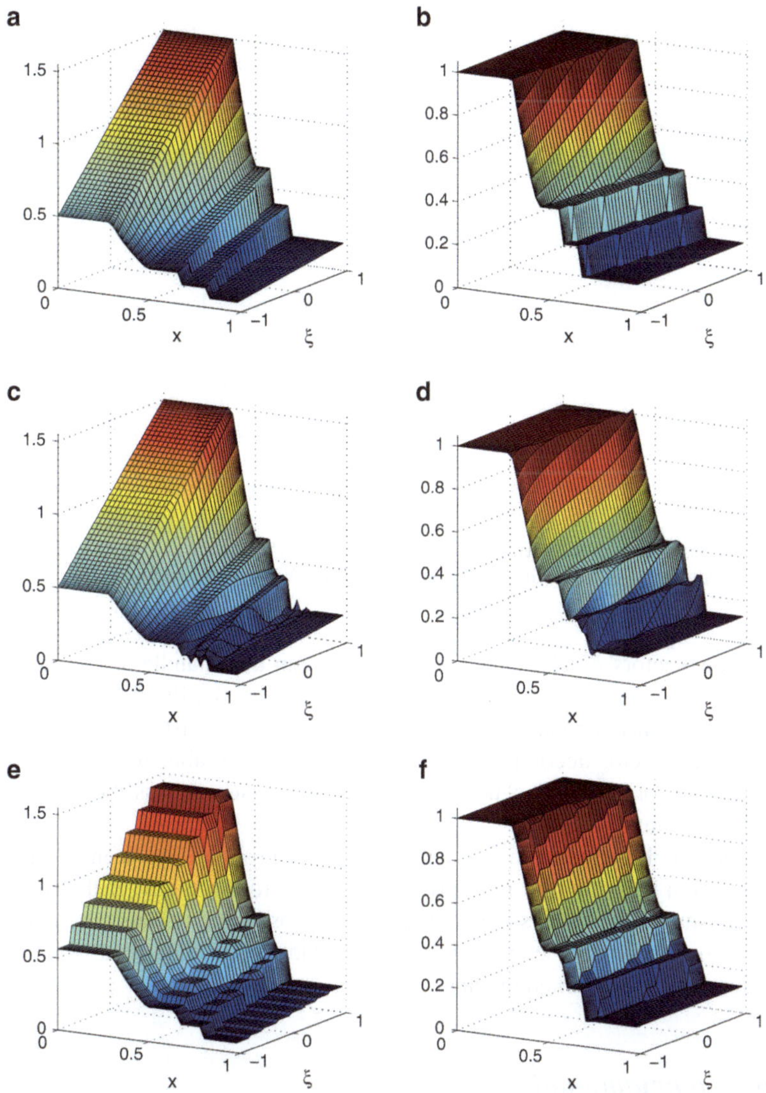

Fig. 8.8 Density as a function of x and ξ at $t = 0.15$. (**a**) Exact solution, case 1. (**b**) Exact solution, case 2. (**c**) Legendre polynomials $(N_p, N_r) = (8, 0)$, case 1. (**d**) Legendre polynomials $(N_p, N_r) = (8, 0)$, case 2. (**e**) Haar wavelets $(N_p, N_r) = (0, 3)$, case 1. (**f**) Haar wavelets $(N_p, N_r) = (0, 3)$, case 2

number to verify that the cases solved for were reasonably close to the supersonic range they model. For this problem, the conservative variable formulation was unstable due to ill-conditioning of the pseudospectral operations except for the original subsonic case 2 ($\rho_R = 0.125$). Note that this numerical breakdown should not be confused with time instability – using analytical decomposition

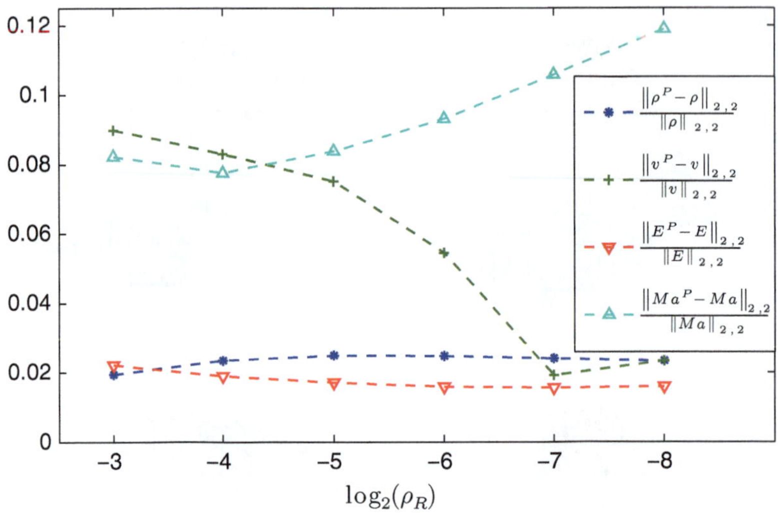

Fig. 8.9 Relative error in density, velocity, energy and Mach number at $t = 0.15$ for different shock strengths. $m = 300$ spatial points, 8 Haar wavelets ($N_p = 0$, $N_r = 3$)

of the eigenvectors of $\boldsymbol{A}$ defined in (8.3), we can also handle supersonic cases with conservative variables. We have not observed any significant variation in the stability properties depending on the order M of the stochastic basis when using constant eigenvector decompositions. Thus, the Roe variable formulation seems more suitable for problems where robustness is an issue, unless the eigenvectors of $\boldsymbol{A}$ are constant.

Legendre polynomials are not suitable for this problem. As seen in Fig. 8.8c, d, the solution is oscillatory in the right state close to the shock. If the right state density is small, as in this supersonic case, such oscillations cause the density to be very close to zero, or even negative. This leads to an unphysical solution and breakdown of the numerical method.

8.3.6 Computational Cost

For stochastic basis functions that admit an eigenvalue decomposition of the matrix $\boldsymbol{A}$ in (8.3) with constant eigenvectors, the computational cost is greatly reduced compared to the general case of non-constant eigenvectors. The Roe average matrices are computed by a series of matrix-vector multiplications only, both for the Roe variables and the conservative variables. The nonlinear pseudospectral operations are also simplified. For instance, the pseudospectral inverse used with the conservative variables can be computed by a series of scalar inverses and matrix-vector multiplications. Let $\boldsymbol{Q}$ be the matrix of constant eigenvectors of $\boldsymbol{A}(.)$. Starting from the gPC expansion ρ^M of the density, put $\rho^{-*} = \boldsymbol{Q}\rho_{EV}^{-*}$, where the vector ρ_{EV}^{-*} is defined by $(\rho_{EV}^{-*})_j = 1/(\sqrt{M}\boldsymbol{Q}^T\rho)_j$ for $j = 1, \ldots, M$. To see that this holds, note that $\Lambda_\rho^M = \sqrt{M} diag(\boldsymbol{Q}^T\rho^M)$. (Superscript M denotes an index, not a power.) Then, with $\mathbf{1}^M = (1, \ldots, 1)^T$ and $e_1 = (1, 0, \ldots, 0)^T$,

$$A(\rho^M)\rho^{-*} = \boldsymbol{Q}\Lambda_\rho^M\boldsymbol{Q}^T\boldsymbol{Q}\rho_{EV}^{-*} = \boldsymbol{Q}\Lambda_\rho^M\rho_{EV}^{-*} = \boldsymbol{Q}\mathbf{1}^M = e_1,$$

so ρ^{-*} has the desired properties of the pseudospectral inverse.

For two stochastic Galerkin systems of order $M = (N_p+1)2^{N_r}$ and $M' = (N_p'+1)2^{N_r'}$ where $M = M'$ but $N_p \neq N_p'$, $N_r' \neq N_r$, the size of the problem and the computational cost are the same. Although the different bases could possibly result in properties that make them very different in the number of iterations required to solve the nonlinear matrix problems, no such tendency was observed. The numerical experiments yield very similar computational costs for the cases tested.

In order to compare the computational cost of the Roe variable expansion with that of the conservative expansion, a similar experimental setup is used for both methods. Sufficiently small test cases are run in order not to exceed the cache limit, which would slow down the simulation time for fine meshes and bias the result. We used test case 1 for short simulation times. Results are shown for both the numerical methods where we use knowledge of the constant eigenvectors of the eigenvalue decomposition of $\boldsymbol{A}$, and the methods designed for the more general case of varying eigenvectors where we have to rely on methods for nonlinear systems.

In the experiments, the same time-step has been used for the different variable expansions, although a larger time-step could be used for the Roe variables. Table 8.1 displays the relative simulation time of the two different variable expansions for an increasing number of Haar wavelets ($M = 2^{N_r}$, $N_p = 0$). One time unit is defined as the time for the numerical simulation of a single deterministic problem using the same numerical method with similar input conditions, discretization and time-step. In the general setup, the higher computational cost for the conservative variable formulation is due to the need to compute inverse quantities and cubic spectral products. The Roe variable formulation only requires solution of the nonlinear system for the square root of the density and quadratic flux function evaluations. The relative benefit of the Roe variable expansion decreases with the

Table 8.1 Relative simulation time using conservative variables and Roe variables, respectively. One time unit is defined as the simulation time of a single deterministic problem with the same time-step as for the MW cases. Results are included for the codes designed for constant eigenvectors and for the same problem with the more general code which does not rely on the assumption of constant eigenvectors

Order of MW	$M = 2$	$M = 4$	$M = 8$	$M = 16$
Time Roe variables, general	29	32	44	107
Time conservative variables, general	267	280	388	60
Time Roe variables, constant eigenvectors	6	7	12	29
Time conservative variables, constant eigenvectors	6	8	13	29

order of wavelet expansion. This is due to the increasing cost of forming spectral products that dominates the total cost for high-order expansions. Note that the difference in computational cost is too large to be due only to the fact that additional pseudospectral operations are required for the conservative variables. The difference is attributed to the fact that the ill-conditioned pseudo-spectral operations may need a large number of iterations.

Using the constant eigenvectors of the eigenvalue decomposition of $\boldsymbol{A}$, we see in Table 8.1 that the two formulations are essentially equivalent since they both reduce to a comparable number of matrix-vector products instead of the solution of nonlinear and possibly ill-conditioned systems. Note that we have not taken into account that the Roe variables permit a time-step larger than that of the conservative variables.

8.4 Summary and Conclusions

A qualitative difference between stochastic Galerkin formulations and non-intrusive formulations is that the numerical analysis of the latter is essentially equivalent to that of deterministic problems, whereas numerical analysis of the former is very different. For the Euler equations, different choice of variables results in different numerical properties of the discretized problem. The stochastic Galerkin counterpart of a simple scalar division in the deterministic or non-intrusive setting may be a potentially ill-conditioned system of equations. However, with careful analysis and suitable solution techniques, the single solution of the more complex stochastic Galerkin problem may be faster than the repeated solutions using a non-intrusive method. The message here is that there is more than one possible stochastic Galerkin formulation, and the increased complexity compared to non-intrusive methods leads to a wider span of numerical performance depending on the choice of numerical solver. In this Chapter, we have compared two formulations of the Euler equations that would not differ in the non-intrusive setting, but behave very differently in the stochastic Galerkin setting.

In computational fluid dynamics, Roe average matrices are used to define averages between neighboring grid cells. The classical example is for the deterministic Euler equations, but for general systems, Roe average matrices are difficult to find. In this Chapter, a Roe average matrix for the standard MUSCL-Roe scheme with Roe variables is derived, and we prove that it satisfies the conditions stated by Roe.

Efficient representation of the input parameters should not be the primary focus in the choice of stochastic basis for hyperbolic problems. Robustness properties over time are far more important, as demonstrated by the test cases of this Chapter. The Legendre polynomial basis exactly represents the input uncertainty in our first test case, but it leads to oscillations around the discontinuity in stochastic space. On the other hand, the Haar wavelets of low-order do not represent the input uncertainty exactly in either test case, but are more robust to discontinuities. Since the optimal stochastic representation is unknown and varies over time, the polynomial chaos framework may be complemented by adaptive methods that adds and removes basis functions over time. These methods will not be further discussed here, but are a natural next step for readers who wish to tackle more complex problems.

The Roe variable formulation is robust for supersonic problems where the conservative variable formulation fails, but only for localized basis functions of the generalized chaos representation. For global Legendre polynomials, the discontinuities in stochastic space lead to oscillations and unphysical behavior of the solution and numerical breakdown. Haar wavelets are more robust in this respect, and do not yield oscillations around discontinuities in stochastic space. The robustness properties can be significantly improved with a stochastic basis that admits an eigenvalue decomposition with constant eigenvectors of the inner triple product matrix that occurs frequently in the evaluation of pseudospectral operations. When this is the case, the Roe variables and conservative variables are similar in performance using the same time-step.

For the general case where we do not assume an eigenvalue decomposition with constant eigenvectors, the Roe variable formulation leads to speedup compared to the conservative variable formulation. The relative speedup decreases with the order of generalized chaos since the total computational cost for high-order expansions is no longer dominated by spectral inversion and square root calculations. Instead, the main cost lies in the formation of spectral product matrices. However, for low-order multiwavelet expansions, the speedup is significant. The difference in computational time is mainly due to the pseudospectral operations of the numerical flux functions, especially if these are ill-conditioned.

We demonstrate the need for robust flux functions by presenting cases where the standard MUSCL-Roe flux fails to capture the solution. The design of a robust numerical method is also highly dependent on the choice of the stochastic basis. The Haar wavelets are not only more robust than Legendre polynomials for representation of discontinuities in stochastic space, but also admit the proof of existence of a Roe matrix and, more specifically, the hyperbolicity of the stochastic Galerkin formulation. This implies that the truncated problem mimics the original problem – a desirable feature.

If the representation of the initial function has not converged, the solution at future times cannot be accurate. The test case with uncertain initial shock location (case 2 in Sect. 8.3.2) illustrates the need to find a representation of uncertainty with fast decay of the coefficients of the generalized chaos expansion. An alternative to more accurate representation of the input uncertainty is to combine the intrusive Roe variable formulation presented here with multielement methods, for instance in the manner presented in [13] or using adaptive methods [12].

References

1. Debusschere BJ, Najm HN, Pébay PP, Knio OM, Ghanem RG, Le Maître OP (2005) Numerical challenges in the use of polynomial chaos representations for stochastic processes. SIAM J Sci Comput 26:698–719. doi:http://dx.doi.org/10.1137/S1064827503427741
2. Le Maître OP, Knio OM (2010) Spectral methods for uncertainty quantification, 1st edn. Springer, Berlin/Heidelberg
3. Le Maître OP, Najm HN, Ghanem RG, Knio OM (2004) Multi-resolution analysis of Wiener-type uncertainty propagation schemes. J Comput Phys 197:502–531. doi:10.1016/j.jcp.2003.12.020, http://portal.acm.org/citation.cfm?id=1017254.1017259
4. LeVeque RJ (2002) Finite volume methods for hyperbolic problems. Cambridge University Press, Cambridge
5. Pettersson P, Abbas Q, Iaccarino G, Nordström J (2009) Efficiency of shock capturing schemes for Burgers' equation with boundary uncertainty. In: Enumath 2009, the eighth European conference on numerical mathematics and advanced applications, Uppsala, June 29–July 3
6. Pettersson P, Iaccarino G, Nordström J (2014) A stochastic Galerkin method for the Euler equations with Roe variable transformation. J Comput Phys 257, Part A(0):481–500. doi:http://dx.doi.org/10.1016/j.jcp.2013.10.011
7. Poëtte G, Després B, Lucor D (2009) Uncertainty quantification for systems of conservation laws. J Comput Phys 228:2443–2467. doi:10.1016/j.jcp.2008.12.018, http://portal.acm.org/citation.cfm?id=1508315.1508373
8. Powell MJD (1970) A Fortran subroutine for solving systems of nonlinear algebraic equations. In: Rabinowitz P (ed) Numerical methods for nonlinear algebraic equations, chap. 7. Gordon and Breach Science Publishers, London/New York
9. Roache PJ (1988) Verification of codes and calculations. AIAA J 36(5):696–702
10. Roe PL (1981) Approximate Riemann solvers, parameter vectors, and difference schemes. J Comput Phys 43(2):357–372. doi:10.1016/0021-9991(81)90128-5, http://www.sciencedirect.com/science/article/B6WHY-4DD1MT3-6G/2/d95f5f5f3b2f002fe5d1fee93f0c6cf8
11. Shunn L, Ham FE, Moin P (2012) Verification of variable-density flow solvers using manufactured solutions. J Comput Phys 231(9):3801–3827
12. Tryoen J, Le Maître OP, Ern A (2012) Adaptive anisotropic spectral stochastic methods for uncertain scalar conservation laws. SIAM J Sci Comput 34(5):A2459–A2481
13. Tryoen J, Le Maître OP, Ndjinga M, Ern A (2010) Intrusive Galerkin methods with upwinding for uncertain nonlinear hyperbolic systems. J Comput Phys 229(18):6485–6511. doi:10.1016/j.jcp.2010.05.007, http://www.sciencedirect.com/science/article/pii/S0021999110002688
14. Tryoen J, Le Maître OP, Ndjinga M, Ern A (2010) Roe solver with entropy corrector for uncertain hyperbolic systems. J Comput Appl Math 235:491–506. doi:http://dx.doi.org/10.1016/j.cam.2010.05.043
15. van Leer B (1979) Towards the ultimate conservative difference scheme. V – a second-order sequel to Godunov's method. J Comput Phys 32:101–136. doi:10.1016/0021-9991(79)90145-1
16. Wan X, Karniadakis GE (2006) Long-term behavior of polynomial chaos in stochastic flow simulations. Comput Methods Appl Math Eng 195:5582–5596

Chapter 9
A Hybrid Scheme for Two-Phase Flow

In this chapter, we investigate a two-phase flow generalization of the Euler equations. A stochastic two-phase problem in one spatial dimension is investigated as a first step towards developing an intrusive method for complex multiphysics problems, such as shock-bubble interactions, high-speed reacting flows with liquid fuels, and Richtmyer-Meshkov instability, with generic with generic uncertainty in the input parameters. So et al. [20] investigated a two-dimensional two-phase problem subject to uncertainty in bubble deformation and contamination of the gas bubble, based on the experiments in [10]. The eccentricity of the elliptic bubble and the ratio of air-helium of the bubble were assumed to be random variables, and quantities of interest were obtained by numerical integration in the stochastic range (stochastic collocation). Previous work on uncertainty quantification for multiphase problems include petroleum reservoir simulations with stochastic point collocation where deterministic flow solvers are evaluated at stochastic collocation points [14] and Karhunen-Loève (KL) expansions combined with perturbation methods [3]. This chapter is based on the work in [17].

We assume uncertainty in the location of the material interface, which requires a stochastic representation of all flow variables. Stochastic quantities are represented as a generalized chaos series, that could be either global as in the case of generalized polynomial chaos [24], or localized (see e.g. [5]). For robustness, we use a generalized chaos expansion with multiwavelets to represent the solution in the stochastic dimension [18]. Note that this basis is global, so the method is fully intrusive. However, the basis is hierarchically localized in the sense that multiwavelets belonging to the same resolution level are grouped into families with non-overlapping support. These features make it suitable for approximating discontinuities in the stochastic space without the oscillations that occur in global polynomial bases.

The stochastic Galerkin method is applied to the stochastic two-phase formulation, resulting in a finite-dimensional deterministic system that shares many

M.P. Pettersson et al., *Polynomial Chaos Methods for Hyperbolic Partial Differential Equations*, Mathematical Engineering,
DOI 10.1007/978-3-319-10714-1_9

properties with the original deterministic problem. The regularity properties of the stochastic problem are essential in the design of an appropriate numerical method. Chen et al. studied the steady-state inviscid Burgers' equation with a source term [4]. We used a similar approach for the inviscid Burgers' equation with uncertain boundary conditions and also analyzed the regularity of low-order stochastic Galerkin approximations of the problem [16]. Schwab and Tokareva analyzed regularity of scalar hyperbolic conservation laws and a linearized version of the Euler equations with uncertain initial profile [19]. In this chapter, we analyze smoothness of the stochastic two-phase problem.

The stochastic Galerkin problem is hyperbolic. This generalized and extended two-phase problem is solved with a hybrid method coupling the continuous phase region with the discontinuous phase region through a numerical interface. The non-smooth region is solved with the HLL-flux, MUSCL-reconstruction in space, and fourth-order Runge-Kutta integration in time. The minmod flux limiter is employed in the experimental results displayed below.

Finite-difference operators in summation-by-parts (SBP) form are used for the high-order spatial discretization. A symmetrized problem formulation that generalizes the energy estimates in [8] for the Euler equations is used for the stochastic Galerkin system. The coupling between the different solution regions is performed with a weak imposition of the interface conditions through an interface using a penalty technique [2]. A fourth-order Runge-Kutta method is used for the integration in time.

9.1 Two-Phase Flow Problem

We assume two phases with volume fractions α and $\beta = 1 - \alpha$ on the domain $x \in [0, 1]$, governed by the advection equation

$$\frac{\partial}{\partial t}\alpha + v'(x,t)\frac{\partial}{\partial x}\alpha = 0, \tag{9.1}$$

where we let $v'(x,t) = v(x,t)$ be the advective velocity obtained from the conservative Euler system below. The Euler equations determine the conservation of masses $\alpha\rho_\alpha$ and $\beta\rho_\beta$, momentum ρv, and total energy E of the two phases through

$$\frac{\partial \boldsymbol{u}}{\partial t} + \frac{\partial \boldsymbol{f}}{\partial x} = 0, \tag{9.2}$$

where

$$\boldsymbol{u} = \begin{bmatrix} \alpha\rho_\alpha \\ \beta\rho_\beta \\ \rho v \\ E \end{bmatrix}, \quad \boldsymbol{f} = \begin{bmatrix} \alpha\rho_\alpha v \\ \beta\rho_\beta v \\ \rho v^2 + p \\ (E + p)v \end{bmatrix}. \tag{9.3}$$

We assume that the pressure p is given by the perfect gas equation of state for two phases

$$p = (\gamma - 1)\left(E - \frac{1}{2}\rho v^2\right), \quad \gamma = \frac{1}{\frac{\alpha}{\gamma_\alpha} + \frac{\beta}{\gamma_\beta}},$$

where γ is the weighted ratio of specific heats. The total density is given by $\rho = \alpha\rho_\alpha + \beta\rho_\beta$. Note that the sum of the first and second equations of (9.2) is the standard mass conservation of the Euler equations. Thus, an equivalent formulation is the Euler equations supplemented with an extra mass conservation equation for one of the phases α and β.

We investigate the Riemann problem defined by the initial conditions

$$(\alpha,\ \alpha\rho_\alpha,\ \beta\rho_\beta,\ \rho v,\ E)^T = \begin{cases} (1,\ 1,\ 0,\ 0,\ 2.5)^T & x < x_0 + \xi \\ (0,\ 0,\ 0.125,\ 0,\ 0.25)^T & x > x_0 + \xi \end{cases}, \tag{9.4}$$

where ξ is a parametrization of the measured or modeled uncertainty in the initial membrane location. Despite the seemingly simple nature of the initial condition, the MW series of the initial condition has an infinite number of nonzero terms. Thus, stochastic truncation error is an issue already at $t = 0$.

The stochastic Galerkin formulation of the two-phase problem is obtained by multiplying (9.1) and (9.2) by each one of the basis functions $\psi_i(\xi)$, and integrating with respect to the probability measure $\mathscr{P}$ over the range of ξ. Initial functions are obtained by projection of (9.4) onto the basis functions $\psi_i(\xi)$. The MW expansion is truncated to $M + 1$ terms and we get the systems for the MW coefficients

$$\frac{\partial}{\partial t}\alpha_k + \sum_{i=0}^{M}\sum_{j=0}^{M} v_i \frac{\partial}{\partial x}\alpha_j \langle\psi_i\psi_j\psi_k\rangle = 0, \quad k = 0, \dots, M, \tag{9.5}$$

$$\beta_k = \delta_{k0} - \alpha_k, \quad k = 0, \dots, M, \tag{9.6}$$

and

$$\frac{\partial}{\partial t}\begin{bmatrix} (\alpha\rho_\alpha)_k \\ (\beta\rho_\beta)_k \\ (\rho v)_k \\ E_k \end{bmatrix} + \frac{\partial}{\partial x}\begin{bmatrix} \sum_{i=0}^{M}\sum_{j=0}^{M}(\alpha\rho_\alpha)_i v_j \langle\psi_i\psi_j\psi_k\rangle \\ \sum_{i=0}^{M}\sum_{j=0}^{M}(\beta\rho_\beta)_i v_j \langle\psi_i\psi_j\psi_k\rangle \\ \sum_{i=0}^{M}\sum_{j=0}^{M}(\rho v)_i v_j \langle\psi_i\psi_j\psi_k\rangle + p_k \\ \sum_{i=0}^{M}\sum_{j=0}^{M}(E_i + p_i) v_j \langle\psi_i\psi_j\psi_k\rangle \end{bmatrix} = 0, \quad k = 0, \dots, M. \tag{9.7}$$

MW expansions for pressure can be updated from the MW of the conservative variables, and then be inserted into the fluxes. It is not possible in general to find a Roe-like variable transformation as was done for the single-phase Euler equations in Chap. 8. We use a pseudospectral approximation of high-order stochastic products for the pressure update. In the computation of, for example, the order M product

$z(\xi) = \sum_{k=0}^{M} z_k \psi_k$ of three stochastic variables $a(\xi), b(\xi), c(\xi)$, for $k = 0, \dots, M$, we use the approximation

$$\begin{aligned} z_k &= (a(\xi)b(\xi)c(\xi))_k = \left\langle \left(\sum_{i=0}^{M} a_i \psi_i(\xi)\right) \left(\sum_{j=0}^{M} b_j \psi_j(\xi)\right) \left(\sum_{l=0}^{M} c_l \psi_l(\xi)\right) \psi_k(\xi) \right\rangle \\ &= \sum_{i=0}^{M}\sum_{j=0}^{M}\sum_{l=0}^{M} \langle \psi_i \psi_j \psi_k \psi_l \rangle a_i b_j c_l \approx \sum_{i=0}^{M}\sum_{m=0}^{M} \langle \psi_i \psi_m \psi_k \rangle a_i \underbrace{\sum_{j=0}^{M}\sum_{l=0}^{M} \langle \psi_j \psi_l \psi_m \rangle b_j c_l}_{(bc)_m^M} \\ &\equiv (a * (b * c))_k, \end{aligned} \tag{9.8}$$

where the pseudospectral product $y = a * b$ of order M is defined by

$$y_k^M = (a * b)_k = \sum_{i=0}^{M}\sum_{j=0}^{M} \langle \psi_i \psi_j \psi_k \rangle a_i b_j, \quad k = 0, \dots, M. \tag{9.9}$$

In matrix notation, we can express (9.9) as

$$\boldsymbol{y}^M = \boldsymbol{A}(\boldsymbol{a}^M)\boldsymbol{b}^M, \tag{9.10}$$

where $\boldsymbol{y}^M = (y_0, \dots, y_M)^T$ is the vector of MW coefficients of y and

$$[\boldsymbol{A}(\boldsymbol{a}^M)]_{j+1,k+1} = \sum_{i=0}^{M} \langle \psi_i \psi_j \psi_k \rangle a_i. \tag{9.11}$$

By successively applying (9.10), we obtain approximations of a range of stochastic functions including polynomials, square roots and inverse quantities [6].

For general stochastic basis functions and general choices of the order of generalized chaos, the stochastic volume fractions α and β cannot be guaranteed to be non-negative. In fact, using first-order Legendre polynomial chaos and projecting the initial condition results in $\alpha(x, t = 0, \xi) = \alpha_0(x, t = 0) + \alpha_1(x, t = 0)\xi$, $\xi \in \mathscr{U}[-1, 1]$. This implies that $\alpha(x, t, \xi) < 0$ for some values of x, t, ξ, which is clearly undesirable. However, we only use Legendre polynomials for the convergence test of a smooth problem using the method of manufactured solutions in Sect. 9.4.1. For the fully discontinuous problem, we use Haar wavelets (piecewise constant multiwavelets, $N_p = 0$) for which the initial function always is physical, no matter the order of wavelet expansion. To see this, we rewrite the stochastic advection system (9.5) in matrix-vector notation,

$$\frac{\partial}{\partial t}\boldsymbol{\alpha}^M + \boldsymbol{A}(\boldsymbol{v}^M(x,t))\frac{\partial}{\partial x}\boldsymbol{\alpha}^M = 0,$$

where $\boldsymbol{A}(.)$ is defined by (9.11). Assuming Haar wavelets, the matrix $\boldsymbol{A}(\boldsymbol{v}^M(x,t))$ in the stochastic advection system can be diagonalized with constant eigenvectors y_k, but space- and time-dependent eigenvalues $\lambda_k(x,t)$, for $k = 0,\dots,M$. The stochastic Galerkin advection problem then decouples to a set of scalar advection problems,

$$\frac{\partial}{\partial t}\tilde{\alpha}_k + \lambda_k(x,t)\frac{\partial}{\partial x}\tilde{\alpha}_k = 0, \quad \tilde{\alpha}(x,t=0) = \tilde{\alpha}_k^{init}(x), \quad k = 0,\dots,M, \tag{9.12}$$

where $\tilde{\alpha}_k = y_k^T \alpha$. The solution of the semilinear advection problem (9.12) is $\tilde{\alpha}_k^{init}(r_k(x,t))$ where r_k is defined by $t = \int_{r_k}^{x} \frac{dx'}{\lambda_k(x',t)}$. No matter the exact form of r_k, the solution will never attain values beyond the range of $\tilde{\alpha}_k^{init}$. This implies that the stochastic volume fraction PDE formulation will never yield unphysical values.

9.2 Smoothness Properties of the Solution

9.2.1 Analytical Solution

The exact solutions to (9.1) and (9.2) subject to (9.4) can be determined analytically, and are discontinuous for all times. The advection problem (9.1) with v independent of x and t has the solution

$$\alpha(x,t) = \alpha_0(x - vt),$$

which is to be interpreted in the weak sense here since it is discontinuous for all t when α_0 is chosen to be a step function. The conservation law (9.2) is a straightforward extension of the Sod test case for shock tube problems, and its exact piecewise smooth solution can be found in [21]. The solution consists of five distinct smooth regions (denoted $\boldsymbol{u}_{(L)}$, $\boldsymbol{u}_{(exp)}$, $\boldsymbol{u}_{(2)}$, $\boldsymbol{u}_{(1)}$, and $\boldsymbol{u}_{(R)}$), and the discontinuities may be found at the interfaces between the different regions. Assume that the initial interface location is $x_0^s = x_0 + \xi$ as given in (9.4). We can then express the deterministic solution for any fixed ξ as a piecewise smooth solution, separated by the four spatial points

$$x_1(t,\xi) = x_0 + \xi - \sqrt{\gamma\frac{p_L}{\rho_L}}t \tag{9.13}$$

$$x_2(t,\xi) = x_0 + \xi + \left(v_2 - \sqrt{\gamma\frac{p_2}{\rho_2}}\right)t \tag{9.14}$$

$$x_3(t,\xi) = x_0 + \xi + v_2 t \tag{9.15}$$

$$x_4(t,\xi) = x_0 + \xi + M_s t, \tag{9.16}$$

where M_s is the Mach number of the shock.

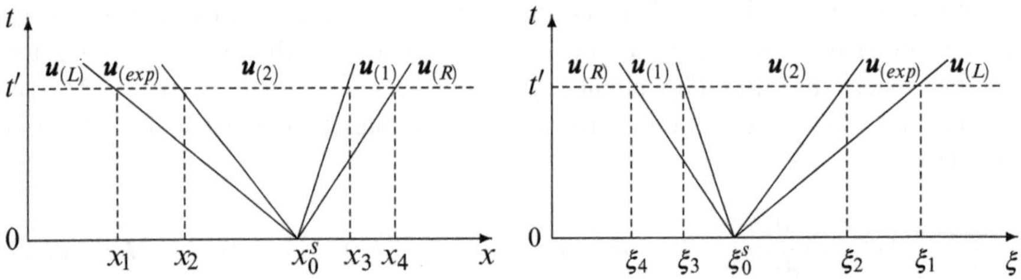

Fig. 9.1 Schematic representation of the solution of the two-phase problem. Solution regions in the $x-t$ space for a fixed ξ (*left*), and solution regions in $\xi - t$ space for a fixed x (*right*)

Any given value of ξ will determine the location of the different regions of piecewise continuous solutions, so the true stochastic solution can be expressed as a function of ξ and the variables of the true deterministic solution. In the x-t-ξ-space, all solution discontinuities are defined by triplets (x, t, ξ) satisfying (9.13)–(9.16). The solution regions are depicted in Fig. 9.1 (left) for any fixed value of ξ.

For any point x, the solution regions can be defined as functions of ξ and t. This is shown in Fig. 9.1 (right), where the points in the stochastic dimension separating the different solution regions are given by

$$\xi_1(x,t) = x - x_0 + \sqrt{\gamma \frac{p_L}{\rho_L}}t \tag{9.17}$$

$$\xi_2(x,t) = x - x_0 - \left(v_2 - \sqrt{\gamma \frac{p_2}{\rho_2}}\right)t \tag{9.18}$$

$$\xi_3(x,t) = x - x_0 - v_2 t \tag{9.19}$$

$$\xi_4(x,t) = x - x_0 - M_s t. \tag{9.20}$$

The solution can be written

$$\begin{aligned} \boldsymbol{u}(x,t,\xi) = \boldsymbol{u}_{(L)} \mathbb{1}_{\{\xi_1<\xi\}} + \boldsymbol{u}_{(exp)}(x-\xi)\mathbb{1}_{\{\xi_2<\xi\leq\xi_1\}} + \boldsymbol{u}_{(2)}\mathbb{1}_{\{\xi_3<\xi\leq\xi_2\}} \\ + \boldsymbol{u}_{(1)}\mathbb{1}_{\{\xi_4<\xi\leq\xi_3\}} + \boldsymbol{u}_{(R)}\mathbb{1}_{\{\xi\leq\xi_{\xi_4}\}}, \end{aligned} \tag{9.21}$$

where the indicator function $\mathbb{1}_{\{A\}}$ of a set A is defined by $\mathbb{1}_{\{A\}}(\xi) = 1$ if $\xi \in A$ and zero otherwise.

Note that if the range of ξ is bounded, some solution states may not occur with nonzero probability for an arbitrary x. The situation shown in Fig. 9.1 (right) requires a sufficiently large range of ξ, or, equivalently, that x is sufficiently close to x_0. The expression (9.21) is always true, however.

9.2.2 Stochastic Modes

The solutions of (9.1) and (9.2) for fixed values of ξ are discontinuous, but the stochastic modes (multiwavelet coefficients) are continuous. To see this, we proceed from the solution (9.21) to derive exact expressions for the stochastic modes. We assume that the probability measure $\mathscr{P}$ has a probability density $\tilde{p}$. The kth mode $\boldsymbol{u}_k$ is given by the projection of (9.21) on $\psi_k(\xi)$,

$$\begin{aligned}\boldsymbol{u}_k(x,t) &= \int_\Omega \boldsymbol{u}(x,t,\xi)\psi_k(\xi)\tilde{p}(\xi)d\xi = \boldsymbol{u}_{(L)}\int_{\xi_1}^{\infty}\psi_k(\xi)\tilde{p}(\xi)d\xi \\ &+ \int_{\xi_2}^{\xi_1}\boldsymbol{u}_{(exp)}(x-\xi)\psi_k(\xi)\tilde{p}(\xi)d\xi + \boldsymbol{u}_{(2)}\int_{\xi_3}^{\xi_2}\psi_k(\xi)\tilde{p}(\xi)d\xi \\ &+ \boldsymbol{u}_{(1)}\int_{\xi_4}^{\xi_3}\psi_k(\xi)\tilde{p}(\xi)d\xi + \boldsymbol{u}_{(R)}\int_{-\infty}^{\xi_4}\psi_k(\xi)\tilde{p}(\xi)d\xi. \end{aligned} \tag{9.22}$$

The density $\tilde{p}$ and multiwavelet ψ_k are at least piecewise continuous functions, so by (9.22) $\boldsymbol{u}_k \in C^0$. Now assume that the parametrization ξ of the uncertainty in the location of x_0 has a probability density $\tilde{p} \in C^s(\mathbb{R})$ for some degree of regularity $s \in \mathbb{N}$. There exists a set $\{\psi_i\}_{i=0}^{\infty}$ of polynomials that are orthogonal with respect to $\tilde{p}$. With this choice of basis functions, we may differentiate (9.22) with respect to x,

$$\begin{aligned}\frac{\partial}{\partial x}\boldsymbol{u}_k &= -\boldsymbol{u}_{(L)}\psi_k(\xi_1)\tilde{p}(\xi_1) + \boldsymbol{u}_{exp}(x-\xi_1)\psi_k(\xi_1)\tilde{p}(\xi_1) - \boldsymbol{u}_{exp}(x-\xi_2)\psi_k(\xi_2)\tilde{p}(\xi_2) \\ &+ \int_{\xi_2}^{\xi_1}\boldsymbol{u}'_{(exp)}(x-\xi)\psi_k(\xi)\tilde{p}(\xi)d\xi + \boldsymbol{u}_{(2)}\psi_k(\xi_2)\tilde{p}(\xi_2) - \boldsymbol{u}_{(2)}\psi_k(\xi_3)\tilde{p}(\xi_3) \\ &+ \boldsymbol{u}_{(1)}\psi_k(\xi_3)\tilde{p}(\xi_3) - \boldsymbol{u}_{(1)}\psi_k(\xi_4)\tilde{p}(\xi_4) + \boldsymbol{u}_{(R)}\psi_k(\xi_4)\tilde{p}(\xi_4), \end{aligned} \tag{9.23}$$

where we used $\partial\xi_i/\partial x = 1$, $i = 1,2,3,4$. In fact, $\boldsymbol{u}_k(x,t)$ as given by (9.22) is $s+1$ times differentiable in x or t for $t > 0$ and $\boldsymbol{u}_k \in C^{s+1}$.

Remark 9.1. Note that the smoothness of $\boldsymbol{u}_k$ in x and t ultimately depends on the smoothness of $\tilde{p}$ and the choice of basis functions $\{\psi_i\}_{i=0}^{\infty}$, which are all functions of ξ. In contrast, for any fixed value of ξ, the solution $\boldsymbol{u}(x,t,\xi)$ is discontinuous in the spatial and temporal dimensions, no matter the smoothness of $\tilde{p}$ and $\{\psi_i\}_{i=0}^{\infty}$.

9.2.3 The Stochastic Galerkin Solution Modes

We investigated the smoothness properties of the stochastic modes of the original problems problem (9.5) and (9.2) above, but in all actual computations we solve the modified stochastic Galerkin approximation (9.5)–(9.7). For low-order MW approximations (small M), the smoothness properties are very different from those derived above. For instance, the $M = 0$ approximation is the deterministic two-phase problem with its characteristic discontinuous solution profile. First-order gPC approximations using a group of orthogonal polynomials and multiwavelets result in linear combinations of deterministic two-phase problems. In terms of regularity, these problems are clearly equivalent to the deterministic problem. Higher-order gPC approximations result in large nonlinear stochastic Galerkin problems that in general cannot be diagonalized into a set of deterministic two-phase problems. Due to their nonlinear nature, we expect these problems to develop discontinuities. However, it is a reasonable assumption that the solution converges to the solution of (9.2). Hence, we assume that the discontinuities get weaker with the order of gPC expansion so that high-order MW approximations have regularity properties that approach the smoothness properties of the analytical stochastic modes.

We have analyzed smoothness of the particular problem of uncertain initial location of the shock in the Riemann problem (9.4). An essential feature of the analysis is that for $t > 0$, the locations of the discontinuities become stochastic. If this were not the case, the gPC coefficients would not be smooth. Thus, for any given set of initial conditions, smoothness should be analyzed in order to determine an appropriate numerical method.

In order to solve (9.5)–(9.7) numerically for arbitrary order M of MW expansion (that may vary in space depending on the smoothness of the solution), we need shock-capturing methods that can account for the discontinuities that are expected due to the stochastic truncation. In regions away from the discontinuities, the solution is at least as smooth as the corresponding deterministic problem, and high-order methods in combination with smooth polynomial stochastic basis functions are more suitable. In the next section, we present a method which combines high-order and shock-capturing methods for the stochastic Galerkin systems.

9.3 Numerical Method

The computational domain is divided into regions of smooth behavior of the solution, and regions of sharp variation. At this stage, these regions are assumed to be known a priori and do not change with time. Thus there is no need to use a detection algorithm to locate the regions of sharp variation apart from flux limiters that are applied for smoothing. However, the methodology may be extended to time-dependent regions (see [7]). A fourth-order Runge-Kutta method is used for the time integration.

9.3.1 Summation-by-Parts Operators

The smooth regions are discretized using a high-order finite difference method based on SBP operators. Boundary conditions are imposed weakly through penalty terms, where the penalty parameters are chosen such that the numerical method is stable. Operators of order $2n$, $n \in \mathbb{N}$, in the interior of the domain are combined with boundary closures of order of accuracy n.

The first derivative SBP operator was introduced in [13, 22]. Let $\vec{u}$ denote the uniform spatial discretization of $\boldsymbol{u}$. For the first derivative, we use the approximation $\boldsymbol{u}_x \approx \boldsymbol{P}^{-1}\boldsymbol{Q}\vec{u}$, where subscript x denotes partial derivative and $\boldsymbol{Q}$ satisfies

$$\boldsymbol{Q} + \boldsymbol{Q}^T = \mathrm{diag}(-1, 0, \ldots, 0, 1) \equiv \tilde{\boldsymbol{B}}. \tag{9.24}$$

The property (9.24) is the almost skew-symmetry property introduced in (4.9). For more details about the SBP framework and use of penalty techniques, see Sect. 4.2.2. As held consistently throughout this book, $\boldsymbol{P}$ must be symmetric and positive definite in order to define a discrete norm. For proof of stability, $\boldsymbol{P}$ must be diagonal.

9.3.2 HLL Riemann Solver

In the non-smooth regions, MUSCL-type flux limiting [23] is used for reconstruction of the left and right states of the conservative fluxes and advection of the volume fractions. For the conservative problem (9.2), we employ the HLL Riemann solver introduced by Harten et al. [12], defined in (4.27). The fastest signal velocities are given by the maximum and minimum eigenvalues of the Jacobian of the flux. In the deterministic case, the eigenvalues of the Jacobian are known analytically, so the method is inexpensive. For the stochastic Galerkin system, analytical expressions are not available, and numerical approximations of the eigenvalues are used instead. In general, obtaining accurate eigenvalue estimates may be computationally costly. However, for the piecewise constant and piecewise linear multiwavelet expansion, we have explicit expressions for the system eigenvalues due to the constant eigenvectors of the inner triple product matrices A given by (9.10), see Appendix B.2.

The HLL-flux and MUSCL reconstruction are applied to solve the conservative problem (9.7). The (standard) MUSCL scheme is used to solve the advection problem (9.5) in the regions where the solution is expected to be non-smooth. In combination with a suitable Runge-Kutta method, the MUSCL scheme is total variation diminshing (TVD) [9]. For the deterministic solution, this would be a sufficient condition for α to attain physically relevant values only. For the stochastic Galerkin system, we need to ensure that the effect of the artificial dissipation from the flux limiters on the different solution modes does not cause the solution α (linear combination of the modes) to become unphysical. By the TVD property, the expectation mode is restricted to $[0, 1]$, so unphysical values can occur only if the high-order modes are less dissipated than the expectation mode. Artificial dissipation affects highly oscillating functions more than slowly oscillating functions. Since the peaks of the initial functions get sharper with the order of wavelet chaos, the higher-order modes are increasingly dissipated by the scheme compared to the lower-order modes and the expectation. Thus, most likely the numerical volume fraction always remains restricted to physically relevant values as time is evolved. This evolution is confirmed by the numerical experiments reported later.

9.3.3 Hybrid Scheme

Numerical interfaces can be designed for stable coupling of problems solved separately using SBP operators. The MUSCL scheme can be rewritten in SBP operator form with an artificial dissipation term [1] and can therefore be coupled with other schemes using SBP operators [7]. The coupling requires the artificial dissipation to be zero at the interface in order to enable energy estimates.

The computational domain is divided into a left smooth solution region and a right non-smooth solution region that are weakly coupled with an interface. The leftmost lying part of the right region is a transition region where a second-order one-sided SBP scheme is applied that transitions into the HLL-MUSCL scheme. In this way, there is a stable coupling between the high-order SBP scheme of the left domain and the second-order SBP scheme of the transition region. Numerical dissipation within the order of the scheme is added to the regions where SBP operators are used. Figure 9.2 schematically depicts the hybrid scheme, applied to two spatial grids and coupled with an interface.

9.3.3.1 An Energy Estimate for the Continuous Problem

We will analyze stability for two solution regions coupled by an interface. However, we start with the *continuous* problem on a single domain. In order to do this, we symmetrize the two-phase problem. We assume the existence of a convex entropy function $S(\boldsymbol{u}^M)$, i.e., the Hessian $\partial^2 S/\partial \boldsymbol{u}_i^M \partial \boldsymbol{u}_j^M$ is positive definite. (Note that convexity as defined here does not allow for zero eigenvalues of the Hessian.) Then, by [11], there exists a variable transformation $\boldsymbol{w}^M(\boldsymbol{u}^M) = \partial S/\partial \boldsymbol{u}^M$ such that $\tilde{\boldsymbol{f}}(\boldsymbol{w}^M) = \boldsymbol{f}(\boldsymbol{u}^M)$ and

$$\tilde{\boldsymbol{H}} \boldsymbol{w}_t^M + \boldsymbol{J}_w \boldsymbol{w}_x^M = 0,$$

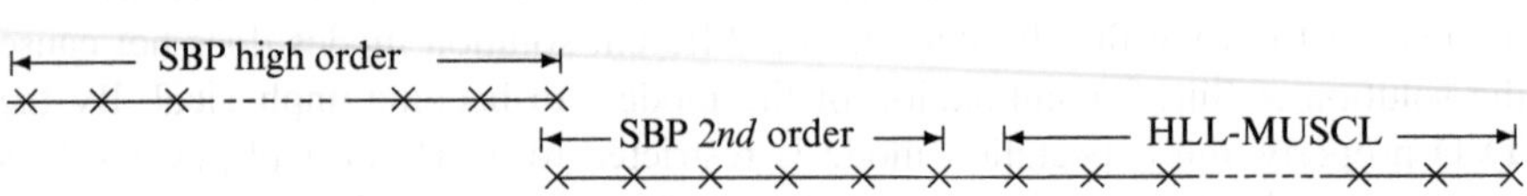

Fig. 9.2 Solution regions on the spatial mesh

where $\boldsymbol{w}^M$ denotes the vector of MW coefficients of the order M approximation of the transformed variables, and the inverse Hessian $\tilde{\boldsymbol{H}} = \partial \boldsymbol{u}^M / \partial \boldsymbol{w}^M = (\partial^2 S / \partial \boldsymbol{u}_i^M \partial \boldsymbol{u}_j^M)^{-1}$ and Jacobian $\boldsymbol{J}_w = \partial \tilde{\boldsymbol{f}} / \partial \boldsymbol{w}$ are symmetric matrices. Due to convexity, $\tilde{\boldsymbol{H}}$ is positive definite and thus defines a norm. As in the case of the Euler equations, the two-phase equations are homogeneous of degree τ, which implies

$$\tilde{\boldsymbol{H}} \boldsymbol{w}^M = \tau \boldsymbol{u}^M \quad \text{and} \quad \boldsymbol{J}_w \boldsymbol{w}^M = \tau \tilde{\boldsymbol{f}}^M. \tag{9.25}$$

We will use the canonical splittings

$$\boldsymbol{u}_t^M = \frac{\tau}{1+\tau} \boldsymbol{u}_t^M + \frac{1}{1+\tau} \tilde{\boldsymbol{H}} \boldsymbol{w}_t^M, \quad \tilde{\boldsymbol{f}}_x^M = \frac{\tau}{1+\tau} \tilde{\boldsymbol{f}}_x^M + \frac{1}{1+\tau} \boldsymbol{J}_w \boldsymbol{w}_x^M.$$

To obtain an energy estimate for the continuous and stability for the semidiscrete problem, the stochastic Galerkin formulation of the two-phase problem must be homogeneous. To show that this holds under the assumption that the corresponding deterministic problem is homogeneous and some additional assumptions, we consider a deterministic problem that is homogeneous of degree τ. Let

$$\boldsymbol{J}(\boldsymbol{u}) \boldsymbol{u} = \tau \boldsymbol{f}(\boldsymbol{u}), \tag{9.26}$$

with solution $\boldsymbol{u} \in \mathbb{R}^n$, Jacobian $\boldsymbol{J} \in \mathbb{R}^{n \times n}$ and flux $\boldsymbol{f} \in \mathbb{R}^n$ for a system of n equations. Now assume that the problem satisfying (9.26) is subject to uncertainty in the parameters or in the input conditions. Let J_{ij} denote the (i, j) entry of $\boldsymbol{J}$ which can be expressed as a truncated MW expansion $J_{ij} = \sum_{k=0}^{M} (J_{ij})_k \psi_k$. The stochastic Galerkin Jacobian $\mathcal{J}^M$ corresponding to $\boldsymbol{J}$ consists of $n \times n$ submatrices, each of size $(M+1) \times (M+1)$. Let $\mathcal{J}_{ij}^M$ be the (i, j) *submatrix* of $\mathcal{J}^M$, defined by

$$[\mathcal{J}_{ij}^M]_{lm} = \langle \psi_l \psi_m J_{ij} \rangle = \sum_{k=0}^{M} (J_{ij})_k \langle \psi_k \psi_l \psi_m \rangle, \; i, j = 1, \ldots, n, \; l, m = 0, \ldots, M. \tag{9.27}$$

The stochastic Galerkin flux vector of MW coefficients $\boldsymbol{f}^M = ((f_1)_0, \ldots, (f_1)_M, \ldots, (f_n)_0, \ldots, (f_n)_M)^T$ is a nonlinear function, and for an arbitrary order M basis of multiwavelets, it is not uniquely defined. To see this, with the pseudospectral product $*$ defined in (9.9), in general

$$(a * b) * c \neq a * (b * c)$$

for MW approximations of stochastic functions $a(\xi), b(\xi), c(\xi)$, each one truncated to some order M. This implies that the definition of the stochastic Galerkin flux $\boldsymbol{f}^M$ depends on the order in which pseudospectral operations are performed when evaluating $\boldsymbol{f}^M$. Hence, it is not uniquely defined. We may now either restrict ourselves to MW bases where the order of pseudospectral operations does not matter e.g., Haar wavelets, or we may restrict the order in which pseudospectral operations are performed so as to make sure that mathematical properties of interest, such as, homogeneity, are satisfied. We take the latter approach and define the order M approximation of $\boldsymbol{f}$ through its MW coefficients by

$$(f_i)_k \equiv \frac{1}{\tau}\sum_{j=1}^{n}(J_{ij} * u_j)_k, \quad i = 1,\dots,n, \quad k = 0,\dots,M, \tag{9.28}$$

which is consistent with the deterministic homogeneous problem. Note that relation (9.28) is essentially just a restriction on the order of pseudospectral operations in the calculation of $\boldsymbol{f}$. It stipulates that $\boldsymbol{f}$ must be defined in terms of the approximation of $\boldsymbol{J}$. Clearly, the approximation of $\boldsymbol{J}$ should also be as close to the true (i.e., infinite order MW expansion) $\boldsymbol{J}$ as possible. However, for the energy estimates that require homogeneity of the stochastic Galerkin formulation, we only need to satisfy (9.28).

Proposition 9.1. *Assume that the deterministic problem (9.26) holds, and for a consistent pseudospectral approximation $\mathscr{J}^M$ of $\boldsymbol{J}$, let the stochastic Galerkin flux $\boldsymbol{f}^M$ be given by the MW coefficients as defined in (9.28). Then the stochastic Galerkin formulation of order M is also homogeneous of degree τ, i.e., it satisfies*

$$\mathscr{J}^M(\boldsymbol{u}^M)\boldsymbol{u}^M = \tau \boldsymbol{f}^M(\boldsymbol{u}^M), \tag{9.29}$$

where $\boldsymbol{u}^M = ((u_1)_0,\dots,(u_1)_M,\dots,(u_n)_0,\dots,(u_n)_M)^T \in \mathbb{R}^{n(M+1)}$ and $\boldsymbol{f}^M = ((f_1)_0,\dots,(f_1)_M,\dots,(f_n)_0,\dots,(f_n)_M)^T \in \mathbb{R}^{n(M+1)}$.

Proof. Using the notation (9.10) for the pseudospectral product $*$, by (9.27) the (i,j) submatrix of $\mathscr{J}^M$ can be written

$$\mathscr{J}^M_{ij} = \boldsymbol{A}\left(\boldsymbol{J}^M_{ij}\right), \quad i,j = 1,\dots,n,$$

where $\boldsymbol{J}^M_{ij} = ((J_{ij})_0,\dots,(J_{ij})_M)^T$. Thus, we have

$$\mathscr{J}^M = \begin{bmatrix} \boldsymbol{A}(\boldsymbol{J}^M_{11}) & \dots & \boldsymbol{A}(\boldsymbol{J}^M_{1n}) \\ \vdots & \ddots & \vdots \\ \boldsymbol{A}(\boldsymbol{J}^M_{n1}) & \dots & \boldsymbol{A}(\boldsymbol{J}^M_{nn}) \end{bmatrix}.$$

By the relation (9.28), any subvector $\boldsymbol{f}^M_i = ((f_i)_0,\dots,(f_i)_M)^T$ of the total flux vector of MW coefficients $\boldsymbol{f}^M$ can be written

$$\boldsymbol{f}^M_i = \frac{1}{\tau}\sum_{j=1}^{n}\boldsymbol{A}(\boldsymbol{J}^M_{ij})\boldsymbol{u}^M_j, \quad i = 1,\dots,n.$$

Then, considering the ith row of submatrices, we have

$$[\mathscr{J}^M \boldsymbol{u}^M]_i = \sum_{j=1}^{n} \mathscr{J}_{ij}^M \boldsymbol{u}_j^M = \sum_{j=1}^{n} \boldsymbol{A}(\boldsymbol{J}_{ij}^M)\boldsymbol{u}_j^M = \tau \boldsymbol{f}_i^M, \quad i = 1, \ldots, n,$$

which is equal to (9.29).

Remark 9.2. The original (deterministic) Jacobian entries J_{ij} are nonlinear functions of u, and the stochastic Galerkin counterpart $\mathscr{J}^M$ is a nonlinear function of the gPC coefficients of u. Since the approximation of a nonlinear stochastic function by means of pseudospectral operations depends on the order in which the operations are performed, the matrix $\mathscr{J}^M$ is not uniquely defined unless we specify the order. However, for proof of Proposition 9.1, it is sufficient to define $\boldsymbol{f}^M$ as a function of $\mathscr{J}^M$, but there is no need to specify $\mathscr{J}^M$ in terms of the order of pseudospectral operations.

We will now derive an energy estimate for the continuous symmetrized formulation of the stochastic Galerkin Euler equations in split form,

$$\frac{\tau}{1+\tau}\boldsymbol{u}_t^M + \frac{1}{1+\tau}\tilde{\boldsymbol{H}}\boldsymbol{w}_t^M + \frac{\tau}{1+\tau}\tilde{\boldsymbol{f}}_x^M + \frac{1}{1+\tau}\boldsymbol{J}_w\boldsymbol{w}_x = 0. \tag{9.30}$$

Under the conditions of Proposition 9.1, multiply (9.30) by $(1+\tau)(\boldsymbol{w}^M)^T$ and integrate over the physical domain. We get

$$\begin{aligned}
\tau \int_0^1 (\boldsymbol{w}^M)^T \boldsymbol{u}_t^M dx &+ \int_0^1 (\boldsymbol{w}^M)^T \tilde{\boldsymbol{H}}\boldsymbol{w}_t^M dx + \tau \int_0^1 (\boldsymbol{w}^M)^T \tilde{\boldsymbol{f}}_x^M dx \\
&+ \int_0^1 (\boldsymbol{w}^M)^T \boldsymbol{J}_w \boldsymbol{w}_x^M dx = \int_0^1 \left((\boldsymbol{w}^M)^T(\tilde{\boldsymbol{H}}\boldsymbol{w}^M)_t + (\boldsymbol{w}^M)^T\tilde{\boldsymbol{H}}\boldsymbol{w}_t^M\right) dx \\
&+ \int_0^1 \left((\boldsymbol{w}^M)^T(\boldsymbol{J}_w\boldsymbol{w}^M)_x + (\boldsymbol{w}^M)^T\boldsymbol{J}_w\boldsymbol{w}_x^M\right) dx \\
&= \frac{d}{dt}\|\boldsymbol{w}^M\|_{\tilde{\boldsymbol{H}}} + [(\boldsymbol{w}^M)^T\boldsymbol{J}_w\boldsymbol{w}^M]_0^1 = 0,
\end{aligned} \tag{9.31}$$

where the first equality follows from (9.25). The generalized energy estimate (9.31) is a straightforward stochastic Galerkin generalization of that given for the deterministic problem in [8].

9.3.3.2 Stability in a Single Domain

Next we consider the *semidiscrete* problem and start with a single domain. The stability analysis is a direct generalization of the stability of the symmetrized Euler equations in [8]. We define the flux and the Jacobian under the conditions of Proposition 9.1 which implies that the stochastic Galerkin system is homogeneous. Let $\vec{\boldsymbol{u}}^M$ and $\vec{\boldsymbol{w}}^M$ denote the spatial discretizations of $\boldsymbol{u}^M$ and $\boldsymbol{w}^M$, respectively, on a mesh consisting of m equidistant gridpoints. Let $\boldsymbol{E}_1 = diag(1, 0, \ldots, 0)$ and $\boldsymbol{E}_m = diag(0, \ldots, 0, 1)$. The semidiscretized scheme is

$$\begin{aligned} &\frac{\tau}{1+\tau}\vec{\boldsymbol{u}}_t^M + \frac{1}{1+\tau}\hat{\boldsymbol{H}}\vec{\boldsymbol{w}}_t^M + \frac{\tau}{1+\tau}(\boldsymbol{P}^{-1}\boldsymbol{Q} \otimes \boldsymbol{I})\tilde{f}^M(\vec{\boldsymbol{w}}^M) \\ &+ \frac{1}{1+\tau}\hat{\boldsymbol{J}}_w(\boldsymbol{P}^{-1}\boldsymbol{Q} \otimes \boldsymbol{I})\vec{\boldsymbol{w}}^M \\ &= (\boldsymbol{P}^{-1}\boldsymbol{E}_1 \otimes \boldsymbol{\Sigma}_1^w)(\vec{\boldsymbol{w}}^M - \vec{\boldsymbol{g}}_1) + (\boldsymbol{P}^{-1}\boldsymbol{E}_m \otimes \boldsymbol{\Sigma}_m^w)(\vec{\boldsymbol{w}}^M - \vec{\boldsymbol{g}}_m), \end{aligned} \tag{9.32}$$

where $\hat{\boldsymbol{H}}$ is block diagonal with each diagonal block equal to $\tilde{\boldsymbol{H}}$ evaluated at the spatial points. $\boldsymbol{\Sigma}_1^w$ and $\boldsymbol{\Sigma}_m^w$ are penalty matrices to be determined, and $\vec{\boldsymbol{g}}_1$ and $\vec{\boldsymbol{g}}_m$ are vectors where only the entries corresponding to the left and right boundaries are allowed nonzero values. We assume a diagonal norm $\boldsymbol{P}$, so $(\boldsymbol{P} \otimes \boldsymbol{I})\hat{\boldsymbol{H}} = \hat{\boldsymbol{H}}(\boldsymbol{P} \otimes \boldsymbol{I})$. Also, $\hat{\boldsymbol{J}}_w$ commutes with $(\boldsymbol{P} \otimes \boldsymbol{I})$. In order to show stability, we may assume homogeneous boundary conditions $\vec{\boldsymbol{g}}_1 = \vec{\boldsymbol{g}}_m = 0$. Multiplying (9.32) from the left by $(1+\tau)(\vec{\boldsymbol{w}}^M)^T(\boldsymbol{P} \otimes \boldsymbol{I})$ and using the homogeneity properties of (9.25) yields

$$\begin{aligned} &\frac{d}{dt}\|\vec{\boldsymbol{w}}^M\|^2_{(\boldsymbol{P}\otimes\boldsymbol{I})\hat{\boldsymbol{H}}} + (\vec{\boldsymbol{w}}^M)^T\left((\boldsymbol{Q} \otimes \boldsymbol{I})\hat{\boldsymbol{J}}_w + \hat{\boldsymbol{J}}_w(\boldsymbol{Q} \otimes \boldsymbol{I})\right)\vec{\boldsymbol{w}}^M \\ &= (1+\tau)(\vec{\boldsymbol{w}}^M)_1^T \boldsymbol{\Sigma}_1^w \vec{\boldsymbol{w}}_1^M + (1+\tau)\vec{\boldsymbol{w}}_m^T \boldsymbol{\Sigma}_m^w \vec{\boldsymbol{w}}_m^M. \end{aligned} \tag{9.33}$$

Add the transpose of (9.33) to itself and use the SBP relation (9.24)

$$\begin{aligned} \frac{d}{dt}\left\|\vec{\boldsymbol{w}}^M\right\|^2_{(\boldsymbol{P}\otimes\boldsymbol{I})\hat{\boldsymbol{H}}} &= \vec{\boldsymbol{w}}_1^T\left(\boldsymbol{J}_w(\vec{\boldsymbol{w}}_1^M) + (1+\tau)\boldsymbol{\Sigma}_1^w\right)\vec{\boldsymbol{w}}_1^M \\ &+ (\vec{\boldsymbol{w}}_m^M)^T\left(-\boldsymbol{J}_w(\vec{\boldsymbol{w}}_m^M) + (1+\tau)\boldsymbol{\Sigma}_m^w\right)\vec{\boldsymbol{w}}_m^M. \end{aligned} \tag{9.34}$$

The scheme is stable in the sense of Definition 1.3 with the penalties

$$\boldsymbol{\Sigma}_1^w = -\delta_1 \boldsymbol{J}_w^+(\vec{\boldsymbol{w}}_1^M), \quad \boldsymbol{\Sigma}_m^w = \delta_m \boldsymbol{J}_w^-(\vec{\boldsymbol{w}}_m^M), \quad \delta_1, \delta_m \geq \frac{1}{1+\tau}.$$

Remark 9.3. The stability analysis above follows that in [8]; for the case $M = 0$ the analysis is in fact identical. We show here that the analysis in [8] generalizes to the stochastic Galerkin formulation of order M of multiwavelet expansion under the conditions of Proposition 9.1.

9.3.3.3 Stability at the Interface

Now consider a problem with two domains connected by an interface. A grid point at the interface will be assigned two solution values, one from each of the stencils that meet at the interface. The difference between the solutions at the interface are penalized analogously to the treatment of the (outer) boundary conditions we have seen in the single domain stability analysis. By ignoring the imposition of boundary conditions, the semidiscrete systems of the left and right domains are given by

$$\frac{\tau}{1+\tau}(\vec{u}_L^M)_t + \frac{1}{1+\tau}\hat{\boldsymbol{H}}(\vec{w}_L^M)_t + \frac{\tau}{1+\tau}(\boldsymbol{P}_L^{-1}\boldsymbol{Q}_L \otimes \boldsymbol{I})\tilde{\boldsymbol{f}}^M(\vec{w}_L^M)$$
$$+\frac{1}{1+\tau}\tilde{\boldsymbol{J}}_w(\boldsymbol{P}_L^{-1}\boldsymbol{Q}_L \otimes \boldsymbol{I})\vec{w}_L^M = (\boldsymbol{P}_L^{-1}\boldsymbol{E}_m \otimes \boldsymbol{\Sigma}_L^w)(\vec{w}_{m,L}^M - \vec{w}_{1,R}^M), \tag{9.35}$$

and

$$\frac{\tau}{1+\tau}(\vec{u}_R^M)_t + \frac{1}{1+\tau}\tilde{\boldsymbol{J}}_u(\vec{w}_R^M)_t + \frac{\tau}{1+\tau}(\boldsymbol{P}_R^{-1}\boldsymbol{Q}_R \otimes \boldsymbol{I})\tilde{\boldsymbol{f}}^M(\vec{w}_R^M)$$
$$+\frac{1}{1+\tau}\tilde{\boldsymbol{J}}_w(\boldsymbol{P}_R^{-1}\boldsymbol{Q}_R \otimes \boldsymbol{I})\vec{w}_R^M = (\boldsymbol{P}_R^{-1}\boldsymbol{E}_1 \otimes \boldsymbol{\Sigma}_R^w)(\vec{w}_{1,R}^M - \vec{w}_{m,L}^M), \tag{9.36}$$

respectively. We follow the procedure of Sect. 9.3.3.2. Multiplying (9.35) from the left by $(1+\tau)(\vec{w}_L^M)^T(\boldsymbol{P}_L \otimes \boldsymbol{I})$ and using the homogeneity identity (9.25), we have

$$\frac{d}{dt}\left\|\vec{w}_L^M\right\|^2_{(\boldsymbol{P}_L\otimes \boldsymbol{I})\hat{\boldsymbol{H}}} + (\vec{w}_L^M)^T(\boldsymbol{Q}_L \otimes \boldsymbol{I})\tilde{\boldsymbol{J}}_w\vec{w}_L^M + (\vec{w}_L^M)^T\tilde{\boldsymbol{J}}_w(\boldsymbol{Q}_L \otimes \boldsymbol{I})\vec{w}_L^M$$
$$= (1+\tau)\vec{w}_{m,L}^M\boldsymbol{\Sigma}_L^w(\vec{w}_{m,L}^M - \vec{w}_{1,R}^M). \tag{9.37}$$

Adding the transpose of (9.37) to itself, neglecting the outer boundaries and performing similar operations on (9.36), we get

$$\frac{d}{dt}\left(\left\|\vec{w}_L^M\right\|^2_{(\boldsymbol{P}_L\otimes \boldsymbol{I})\hat{\boldsymbol{H}}} + \left\|\vec{w}_R^M\right\|^2_{(\boldsymbol{P}_R\otimes \boldsymbol{I})\hat{\boldsymbol{H}}}\right) + (\vec{w}_{m,L}^M)^T\boldsymbol{J}_w(\vec{w}_{m,L}^M)\vec{w}_{m,L}^M$$
$$-(\vec{w}_{1,R}^M)^T\boldsymbol{J}_w(\vec{w}_{1,R}^M)\vec{w}_{1,R}^M = (1+\tau)(\vec{w}_{m,L}^M)^T\boldsymbol{\Sigma}_L^w(\vec{w}_{m,L}^M - \vec{w}_{1,R}^M)$$
$$+(1+\tau)(\vec{w}_{1,R}^M)^T\boldsymbol{\Sigma}_R^w(\vec{w}_{1,R}^M - \vec{w}_{m,L}^M). \tag{9.38}$$

Assuming symmetric $\boldsymbol{\Sigma}_L^w$ and $\boldsymbol{\Sigma}_R^w$, we get the stability condition

$$\begin{bmatrix}\vec{w}_{m,L}^M \\ \vec{w}_{1,R}^M\end{bmatrix}^T\begin{bmatrix}-\boldsymbol{J}_w(\vec{w}_{m,L}^M) + (1+\tau)\boldsymbol{\Sigma}_L^w & -\frac{1+\tau}{2}(\boldsymbol{\Sigma}_L^w + \boldsymbol{\Sigma}_R^w) \\ -\frac{1+\tau}{2}(\boldsymbol{\Sigma}_L^w + \boldsymbol{\Sigma}_R^w) & \boldsymbol{J}_w(\vec{w}_{1,R}^M) + (1+\tau)\boldsymbol{\Sigma}_R^w\end{bmatrix}\begin{bmatrix}\vec{w}_{m,L}^M \\ \vec{w}_{1,R}^M\end{bmatrix} \le 0. \tag{9.39}$$

Being in the smooth domain, we assume $\boldsymbol{J}_{\boldsymbol{w}}(\vec{\boldsymbol{w}}_{m,L}^{M}) = \boldsymbol{J}_{\boldsymbol{w}}(\vec{\boldsymbol{w}}_{1,R}^{M}) = \boldsymbol{J}$ and obtain stability with

$$\boldsymbol{\Sigma}_L^{\boldsymbol{w}} = \frac{1}{1+\tau}\boldsymbol{J} - \boldsymbol{\theta}, \quad \boldsymbol{\Sigma}_R^{\boldsymbol{w}} = -\frac{1}{1+\tau}\boldsymbol{J} - \boldsymbol{\theta},$$

where $\boldsymbol{\theta}$ is a positive semidefinite matrix. This is completely analogous to the penalties derived in the constant advection problem presented in [7].

The penalties derived in the stability analysis apply to the entropy variables $\boldsymbol{w}$, but in the numerical experiments we use a conservative formulation for correct shock speed and employ the conservative variables $\boldsymbol{u}$. Therefore we need to transform the penalties to the conservative variables. Assuming that the solution is smooth and $\tilde{\boldsymbol{H}}(\vec{\boldsymbol{w}}_{m,L}^{M}) = \tilde{\boldsymbol{H}}(\vec{\boldsymbol{w}}_{1,R}^{M})$, we rewrite the interface terms

$$\begin{aligned}
\boldsymbol{\Sigma}_L^{w}(\vec{\boldsymbol{w}}_{m,L}^{M} - \vec{\boldsymbol{w}}_{1,R}^{M}) &= \frac{1}{1+\tau}\boldsymbol{J}(\vec{\boldsymbol{w}}_{m,L}^{M} - \vec{\boldsymbol{w}}_{1,R}^{M}) - \boldsymbol{\theta}(\vec{\boldsymbol{w}}_{m,L}^{M} - \vec{\boldsymbol{w}}_{1,R}^{M}) \\
&= \frac{1}{1+\tau}\left(\tilde{\boldsymbol{f}}^{M}(\vec{\boldsymbol{w}}_{m,L}^{M}) - \tilde{\boldsymbol{f}}^{M}(\vec{\boldsymbol{w}}_{1,R}^{M})\right) \\
&\quad -\boldsymbol{\theta}\tau\left(\tilde{\boldsymbol{H}}^{-1}(\vec{\boldsymbol{w}}_{m,L}^{M})\vec{\boldsymbol{u}}_{m,L}^{M} - \tilde{\boldsymbol{H}}^{-1}(\vec{\boldsymbol{w}}_{1,R}^{M})\vec{\boldsymbol{u}}_{1,R}^{M}\right) \\
&= \frac{1}{1+\tau}\left(\boldsymbol{f}^{M}(\vec{\boldsymbol{u}}_{m,L}^{M}) - \boldsymbol{f}^{M}(\vec{\boldsymbol{u}}_{1,R}^{M})\right) - \hat{\boldsymbol{\theta}}\left(\vec{\boldsymbol{u}}_{m,L}^{M} - \vec{\boldsymbol{u}}_{1,R}^{M}\right) \\
&= \left(\frac{1}{1+\tau}\boldsymbol{J}_{\boldsymbol{u}} - \hat{\boldsymbol{\theta}}\right)\left(\vec{\boldsymbol{u}}_{m,L}^{M} - \vec{\boldsymbol{u}}_{1,R}^{M}\right) = \boldsymbol{\Sigma}_L^{\boldsymbol{u}}\left(\vec{\boldsymbol{u}}_{m,L}^{M} - \vec{\boldsymbol{u}}_{1,R}^{M}\right),
\end{aligned} \tag{9.40}$$

where

$$\boldsymbol{J}_{\boldsymbol{u}} = \left.\frac{\partial \boldsymbol{f}^{M}}{\partial \boldsymbol{u}^{M}}\right|_{x=x_{int}},$$

and

$$\boldsymbol{\Sigma}_L^{\boldsymbol{u}} = \frac{1}{1+\tau}\boldsymbol{J}_{\boldsymbol{u}} - \hat{\boldsymbol{\theta}}, \tag{9.41}$$

and $\hat{\boldsymbol{\theta}} = \tau\boldsymbol{\theta}\tilde{\boldsymbol{H}}^{-1}$ is a positive semidefinite matrix for $\tau > 0$ since $\tilde{\boldsymbol{H}}$ is positive definite and $\boldsymbol{\theta}$ is positive semidefinite. Similarly, we get the right penalty matrix

$$\boldsymbol{\Sigma}_R^{\boldsymbol{u}} = -\frac{1}{1+\tau}\boldsymbol{J}_{\boldsymbol{u}} - \hat{\boldsymbol{\theta}}. \tag{9.42}$$

With the penalty matrices (9.41) and (9.42), we obtain stability, as defined in Definition 1.3.

9.3.3.4 Conservation at the Interface

In order to show conservation over the interface, we mimic the continuous case where we multiply the conservative formulation by a smooth function ϕ, integrate by parts to get

$$\int_0^{x_{int}} \phi u_t dx + \int_{x_{int}}^1 \phi u_t dx = \int_0^{x_{int}} \phi_x f(u) dx + \int_{x_{int}}^1 \phi_x f(u) dx + B.T., \tag{9.43}$$

where $B.T.$ denotes outer boundary terms. In (9.43) no interface terms are present. Consider the semidiscrete scheme

$$(\vec{\boldsymbol{u}}_L^M)_t + (\boldsymbol{P}_L^{-1}\boldsymbol{Q}_L \otimes \boldsymbol{I})\boldsymbol{f}^M(\vec{\boldsymbol{u}}_L^M) = (\boldsymbol{P}_L^{-1}\boldsymbol{E}_m \otimes \boldsymbol{\Sigma}_L^u)(\vec{\boldsymbol{u}}_{m,L}^M - \vec{\boldsymbol{u}}_{1,R}^M) \tag{9.44}$$

$$(\vec{\boldsymbol{u}}_R^M)_t + (\boldsymbol{P}_R^{-1}\boldsymbol{Q}_R \otimes \boldsymbol{I})\boldsymbol{f}^M(\vec{\boldsymbol{u}}_R^M) = (\boldsymbol{P}_R^{-1}\boldsymbol{E}_1 \otimes \boldsymbol{\Sigma}_R^u)(\vec{\boldsymbol{u}}_{1,R}^M - \vec{\boldsymbol{u}}_{m,L}^M). \tag{9.45}$$

Multiplying from the left by $\vec{\boldsymbol{\phi}}_L^T(\boldsymbol{P}_L \otimes \boldsymbol{I})$ and $\vec{\boldsymbol{\phi}}_R^T(\boldsymbol{P}_R \otimes \boldsymbol{I})$, respectively, where $\vec{\boldsymbol{\phi}}_L$ and $\vec{\boldsymbol{\phi}}_L$ are discretized smooth functions satisfying $\vec{\boldsymbol{\phi}}_{m,L} = \vec{\boldsymbol{\phi}}_{1,R} = \vec{\boldsymbol{\phi}}_I$, we get

$$\begin{aligned}
\vec{\boldsymbol{\phi}}_L^T(\boldsymbol{P}_L \otimes \boldsymbol{I})(\vec{\boldsymbol{u}}_L^M)_t + \vec{\boldsymbol{\phi}}_R^T(\boldsymbol{P}_R \otimes \boldsymbol{I})(\vec{\boldsymbol{u}}_R^M)_t &= (\boldsymbol{D}_L\vec{\boldsymbol{\phi}}_L)^T(\boldsymbol{P}_L \otimes \boldsymbol{I})\boldsymbol{f}^M(\vec{\boldsymbol{u}}_L^M) \\
&\quad + (\boldsymbol{D}_R\vec{\boldsymbol{\phi}}_R)^T(\boldsymbol{P}_R \otimes \boldsymbol{I})\boldsymbol{f}^M(\vec{\boldsymbol{u}}_R^M) + B.T. \\
+ \vec{\boldsymbol{\phi}}_I^T \Big[(\vec{\boldsymbol{u}}_{m,L}^M - \vec{\boldsymbol{u}}_{1,R}^M)(\boldsymbol{\Sigma}_L^u - \boldsymbol{\Sigma}_R^u) &- \boldsymbol{f}^M(\vec{\boldsymbol{u}}_{m,L}^M) + \boldsymbol{f}^M(\vec{\boldsymbol{u}}_{1,R}^M)\Big].
\end{aligned} \tag{9.46}$$

The semidiscrete formulation (9.46) mimics the continuous expression (9.43) if we choose $\boldsymbol{\Sigma}_L^u$ and $\boldsymbol{\Sigma}_R^u$ such that

$$(\vec{\boldsymbol{u}}_{m,L}^M - \vec{\boldsymbol{u}}_{1,R}^M)(\boldsymbol{\Sigma}_L^u - \boldsymbol{\Sigma}_R^u) - \boldsymbol{f}^M(\vec{\boldsymbol{u}}_{m,L}^M) + \boldsymbol{f}^M(\vec{\boldsymbol{u}}_{1,R}^M) = \boldsymbol{0}.$$

We assume $\boldsymbol{J}_w(\vec{\boldsymbol{w}}_{m,L}^M) = \boldsymbol{J}_w(\vec{\boldsymbol{w}}_{1,R}^M) = \boldsymbol{J}$. Then, the interface terms cancel with the choice $\boldsymbol{\Sigma}_L^u - \boldsymbol{\Sigma}_R^u = \boldsymbol{J}$, which is consistent with the condition for stability given by the penalties (9.41) and (9.42) and $\tau = 1$.

9.4 Numerical Results

The exact solution of the test problem is known analytically for any given value of the stochastic variable ξ. Thus, we can obtain the exact statistics to arbitrary accuracy by averaging the exact Riemann solutions over a large number of realizations of ξ. In the numerical experiments, we will assume $\xi \sim \mathscr{U}[-0.05, 0.05]$, where $\mathscr{U}$ denotes the uniform distribution. For the numerical solutions, we use SBP operators that can be found in [15].

9.4.1 Convergence of Smooth Solutions

The method of manufactured solutions is used to impose a smooth timedependent solution of the two-phase problem through a source term. We consider the manufactured solution defined by

$$\begin{aligned}
\alpha &= \alpha_0 + \alpha_1 \tanh(s(x_0 - x + t + \xi)) \\
\beta &= \beta_0 + \beta_1 \tanh(-s(x_0 - x + t + \xi)) \\
v &= \tanh(s(v_0 + x_0 - x + t + \xi)) + \tanh(-s(-v_0 + x_0 - x + t + \xi)) \\
p &= p_0 + p_1 \tanh(s(x_0 - x + t + \xi)),
\end{aligned}$$

with $s = 15$, $v_0 = 0.03$, $\alpha_0 = \alpha_1 = \beta_0 = \beta_1 = 0.5$, $p_0 = 0.75$, $p_1 = 0.25$. We take $\rho_\alpha = 1$ and $\rho_\beta = 0.125$. We measure the error in the $L_2(\Omega, \mathscr{P})$ norm and the discrete ℓ_2 norm,

$$\begin{aligned}
\left\| \vec{u}^M - \vec{u} \right\|_{2,2} &\equiv \left\| \vec{u}^M - \vec{u} \right\|_{\ell_2, L_2(\Omega,\mathscr{P})} \\
&= \left(\Delta x \sum_{i=1}^{m} \left\| \vec{u}^M(x_i, t, \xi) - \vec{u}(x_i, t, \xi) \right\|^2_{L_2(\Omega,\mathscr{P})} \right)^{1/2} \\
&= \left(\Delta x \sum_{i=1}^{m} \int_\Omega (\vec{u}^M(x_i, t, \xi) - \vec{u}(x_i, t, \xi))^2 d\mathscr{P}(\xi) \right)^{1/2} \\
&\approx \left(\Delta x \sum_{i=1}^{m} \sum_{j=1}^{q} (\vec{u}^M(x_i, t, \xi_q^{(j)}) - \vec{u}(x_i, t, \xi_q^{(j)}))^2 w_q^{(j)} \right)^{1/2},
\end{aligned} \tag{9.47}$$

where a q-point quadrature rule with points $\{\xi_q^{(j)}\}_{j=1}^q$ and weights $\{w_q^{(j)}\}_{j=1}^q$ was used in the last line to approximate the integral in ξ. The Gauss-Legendre quadrature is used here since the solution is smooth in the stochastic dimension.

Figure 9.3a shows the spatial convergence when the proportion of low-order and high-order points remains constant. The low-order scheme dominates the error, so the overall convergence rate is second-order. In regions of fourth-order operators, the error levels are lower and therefore the local accuracy higher compared to the regions of second-order operators, see Fig. 9.3b. This is further illustrated in

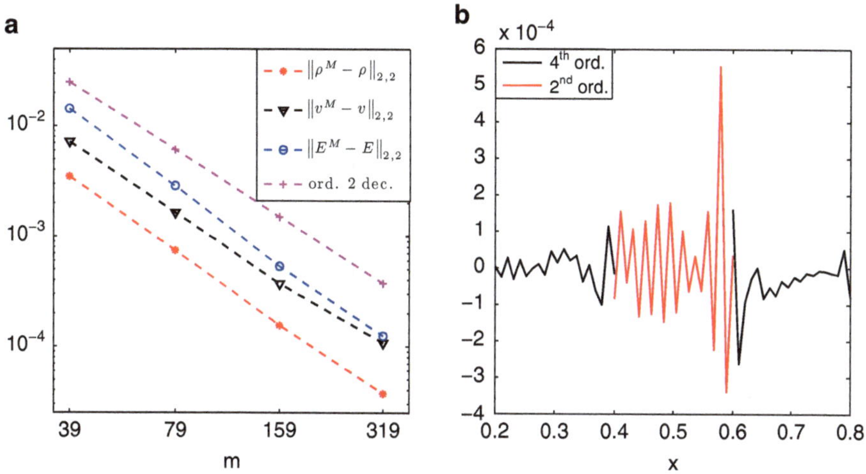

Fig. 9.3 SBP 4-2-4, fixed proportion of SBP 2 points. $N_p = 8$, $N_r = 0$ order of multiwavelets (Legendre polynomials). (**a**) 2,2 norm of errors for smooth solution, $t = 0.05$. (**b**) Error in mean density, $t = 0.1$

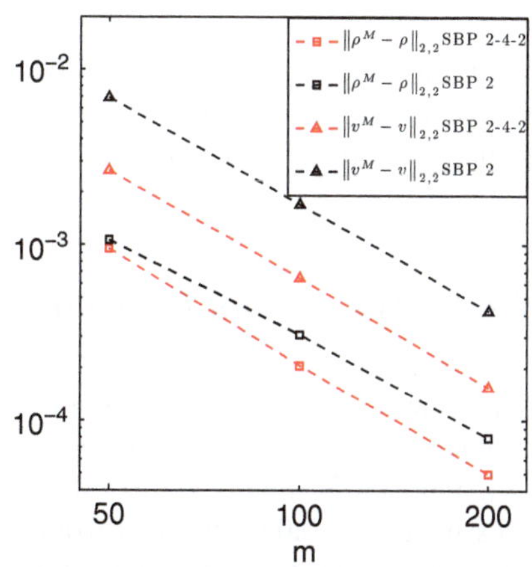

Fig. 9.4 Comparison of 2,2 norm of errors, three solution regions SBP 2-4-2 versus a single region solved with SBP 2, $t = 0.1$. The proportion of fourth-order points remains constant during mesh refinement. $N_p = 8$, $N_r = 0$ order of multiwavelets (Legendre polynomials)

Fig. 9.4, where a similar problem with sharp gradients in the middle of the domain is solved with a hybrid scheme where fourth-order operators are used for the region of large gradients and second-order operators are used for the regions next to the boundaries. With a constant proportion of high-order points under mesh refinement, the convergence is second-order. Comparison with the solution with second-order operators throughout the computational domain, also included in Fig. 9.4, shows that the error of the hybrid scheme is smaller.

Figure 9.5a shows the spatial convergence employing three computational domains separated by two interfaces. The middle domain is solved with second-order SBP and the left and right domains with fourth-order SBP. The number of points in the second-order region remains constant (20), as the high-order domains

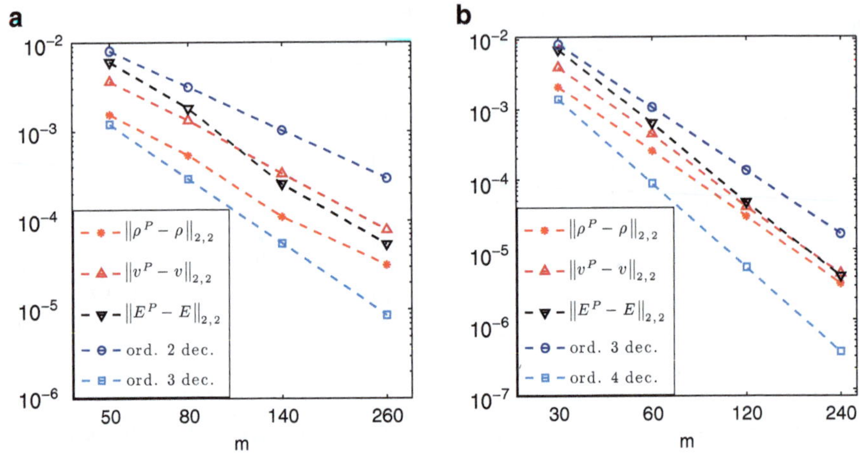

Fig. 9.5 Spatial convergence with three regions and two interfaces. $t = 0.05$, $N_p = 8$, $N_r = 0$ order of multiwavelets (Legendre polynomials). Superscript P denotes the numerical gPC solution. (**a**) SBP4-SBP2-SBP4, fixed number of SBP2 points. (**b**) Three SBP4 schemes coupled by two interfaces

are refined. Figure 9.5b depicts the spatial convergence with three domains, all solved with fourth-order SBP. The proportion of points in each region remains the same, so the interface locations do not change when the grids are refined.

9.4.2 Non-smooth Riemann Problem

With the hybrid scheme as depicted schematically in Fig. 9.2, we solve the problems (9.5)–(9.7) with the boundary conditions in (9.4) and assuming $\xi \sim \mathscr{U}[-0.05, 0.05]$. Figure 9.6 shows the variances of density, velocity, energy, and pressure at $t = 0.05$. The error from the interface is not significant compared to the error due to the stochastic truncation and spatial resolution. A relatively fine mesh and high-order MW expansion is required to capture the variance of the solution. Especially high-order MW coefficients exhibit sharp spatial variation. Thus, to attain a given level of accuracy, more spatial gridpoints are required for the stochastic Galerkin problem compared to the deterministic problem.

Figure 9.7 depicts the convergence of pressure statistics with increasing order of MW on a fixed spatial grid of 400 points. In the analysis of regularity in Sect. 9.2, we anticipated the solution to develop a larger number of weaker discontinuities as the order of MW expansion increases. This behavior can be observed in Fig. 9.7. All (visible) discontinuities are located in the right domain where the shock-capturing method is used.

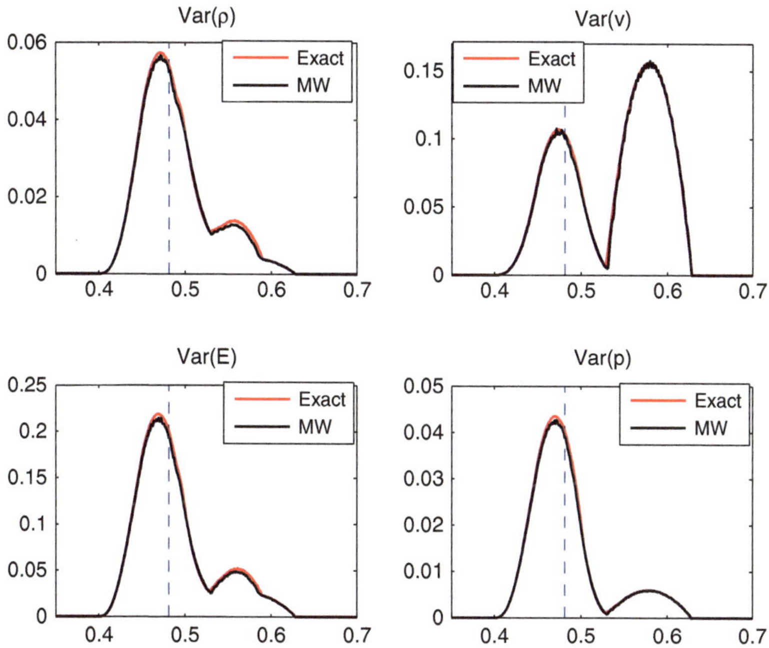

Fig. 9.6 Variances at $t = 0.05$, $m = 400$, fourth order SBP (*left region*) single interface (*dashed blue line*) and HLL-MUSCL (*right region*), $(N_p, N_r) = (0, 5)$ (Haar wavelets)

9.5 Summary and Conclusions

In order to efficiently solve fluid flow problems, a feasible strategy is to locally adapt the numerical method to the smoothness of the solution whenever these properties are known or can be estimated. Stochastic Galerkin formulation of a stochastic hyperbolic problem typically leads to a problem that develops multiple discontinuities in finite time. If these discontinuities are all contained within a known spatial region in a time interval of interest, a shock-capturing method should be used for the corresponding grid points. For the regions of smooth solution, high-order methods should be used. The different methods must be coupled to maintain stability and propagate information accurately over the interfaces between the domains. Note that the solution is unknown in the interior, so one cannot treat the interfaces like boundaries with known boundary conditions. A two-phase Riemann problem with uncertain initial discontinuity location has been investigated with respect to the smoothness properties of the MW coefficients of the solution. Whereas the corresponding deterministic problem has a discontinuous solution profile, the stochastic modes of the gPC expansion of the true solution are smooth.

A symmetrization and combination of conservative and non-conservative formulation leads to a generalized energy estimate for the stochastic Galerkin system, just as for the case of the deterministic Euler equations. Under certain smoothness

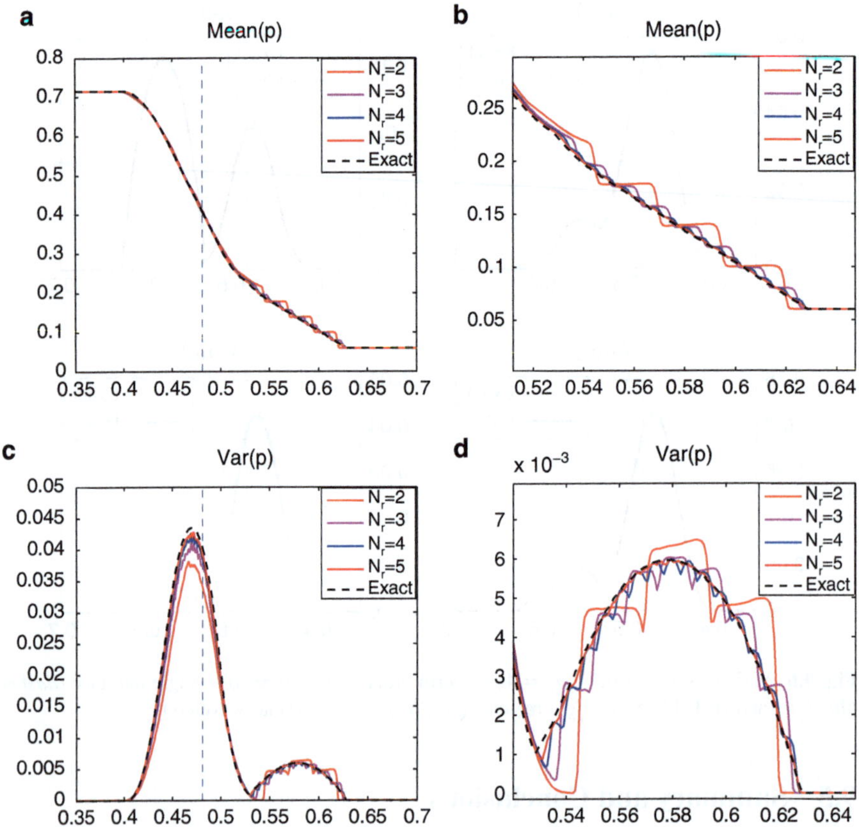

Fig. 9.7 Convergence of the mean and variance of pressure with the order of MW chaos, different orders of piecewise constant MW. $t = 0.05$, $m = 400$. Fourth-order SBP (left domain) and HLL-MUSCL (right domain). (**a**) Mean pressure. (**b**) Mean pressure in the proximity of the deterministic shock. (**c**) Variance of pressure. (**d**) Variance of pressure in the proximity of the deterministic shock

assumptions, stability at the interfaces can be obtained for the symmetrized system. The derived penalty matrices are transformed back to the conservative variable formulation that is used in the numerical experiments.

The numerical results show that the convergence rate for the smooth problem (smoothness enforced by the method of manufactured solutions) is second-order when fourth-order and second-order operators are combined and the proportion of second-order points remains constant during mesh refinement. However, the error is smaller in this case compared to the case of a single domain solved with second-order operators.

The two-phase non-smooth Riemann problem is reasonably well resolved with the hybrid scheme combining high-order SBP operators in the smooth regions with the HLL solver and MUSCL reconstruction in the spatial region containing discontinuities. A relatively large number of multiwavelets is needed to accurately represent the stochastic solution. This in turn requires a fine spatial mesh for accurate resolution.

The framework presented here can be extended to time-dependent interfaces that are adapted to the evolving regions of non-smooth solutions. A moving mesh based on interfaces and SBP-operators has already been designed for deterministic problems in [7], and this technique could be used for stochastic Galerkin systems. Depending on the problem, different MW bases can be used in the different spatial regions for efficient representation of the local uncertainty. Alternative techniques include adaptive stochastic bases that evolve in space and time for optimal representation of localized phenomena.

In the case of a moving interface, several detection strategies can be used relying either on the physical solution, e.g., $\alpha = 0.5$, or on the overall variance which would provide a measure of the oscillations. Further investigations are required to identify the most effective detection algorithm.

References

1. Abbas Q, van der Weide E, Nordström J (2009) Accurate and stable calculations involving shocks using a new hybrid scheme. In: Proceedings of the 19th AIAA CFD conference, no. 2009–3985. Conference Proceeding Series. AIAA, San Antonio, TX.
2. Carpenter MH, Nordström J, Gottlieb D (1999) A stable and conservative interface treatment of arbitrary spatial accuracy. J Comput Phys 148(2):341–365. doi:http://dx.doi.org/10.1006/jcph.1998.6114
3. Chen M, Keller AA, Lu Z, Zhang D, Zyvoloski G (2008) Uncertainty quantification for multiphase flow in random porous media using KL-based moment equation approaches. In: AGU spring meeting abstracts, p A2, Fort Lauderdale, FL.
4. Chen QY, Gottlieb D, Hesthaven JS (2005) Uncertainty analysis for the steady-state flows in a dual throat nozzle. J Comput Phys 204(1):378–398. doi:http://dx.doi.org/10.1016/j.jcp.2004.10.019
5. Deb MK, Babuška IM, Oden JT (2001) Solution of stochastic partial differential equations using Galerkin finite element techniques. Comput Methods Appl Math 190(48):6359–6372. doi:10.1016/S0045-7825(01)00237-7, http://www.sciencedirect.com/science/article/pii/S0045782501002377
6. Debusschere BJ, Najm HN, Pébay PP, Knio OM, Ghanem RG, Le Maître OP (2005) Numerical challenges in the use of polynomial chaos representations for stochastic processes. SIAM J Sci Comput 26:698–719. doi:http://dx.doi.org/10.1137/S1064827503427741
7. Eriksson S, Abbas Q, Nordström J (2011) A stable and conservative method for locally adapting the design order of finite difference schemes. J Comput Phys 230(11):4216–4231. doi:http://dx.doi.org/10.1016/j.jcp.2010.11.020
8. Gerritsen M, Olsson P (1996) Designing an efficient solution strategy for fluid flows. J Comput Phys 129(2):245–262. doi:10.1006/jcph.1996.0248, http://dx.doi.org/10.1006/jcph.1996.0248
9. Gottlieb S, Shu CW (1998) Total variation diminishing runge-kutta schemes. Math Comput 67:73–85
10. Haas JF, Sturtevant B (1987) Interaction of weak shock waves with cylindrical and spherical gas inhomogeneities. J Fluid Mech 181:41–76. doi:10.1017/S0022112087002003

11. Harten A (1983) On the symmetric form of systems of conservation laws with entropy. J Comput Phys 49(1):151–164. doi:10.1016/0021-9991(83)90118-3, http://www.sciencedirect.com/science/article/pii/0021999183901183
12. Harten A, Lax PD, van Leer B (1983) On upstream differencing and Godunov-type schemes for hyperbolic conservation laws. SIAM Rev 25(1):35–61. http://www.jstor.org/stable/2030019
13. Kreiss HO, Scherer G (1974) Finite element and finite difference methods for hyperbolic partial differential equations. In: Mathematical aspects of finite elements in partial differential equations. Academic, New York, pp 179–183
14. Li H, Zhang D (2009) Efficient and accurate quantification of uncertainty for multiphase flow with the probabilistic collocation method. SPE J 14(4):665–679
15. Mattsson K, Nordström J (2004) Summation by parts operators for finite difference approximations of second derivatives. J Comput Phys 199(2):503–540. doi:10.1016/j.jcp.2004.03.001, http://dx.doi.org/10.1016/j.jcp.2004.03.001
16. Pettersson P, Iaccarino G, Nordström J (2009) Numerical analysis of the Burgers' equation in the presence of uncertainty. J Comput Phys 228:8394–8412. doi:10.1016/j.jcp.2009.08.012, http://dl.acm.org/citation.cfm?id=1621150.1621394
17. Pettersson P, Iaccarino G, Nordström J (2013) An intrusive hybrid method for discontinuous two-phase flow under uncertainty. Comput Fluids 86(0):228–239. doi:http://dx.doi.org/10.1016/j.compfluid.2013.07.009
18. Pettit CL, Beran PS (2006) Spectral and multiresolution Wiener expansions of oscillatory stochastic processes. J Sound Vib 294:752–779
19. Schwab C, Tokareva SA (2011) High order approximation of probabilistic shock profiles in hyperbolic conservation laws with uncertain initial data. Tech Rep 2011–53, ETH, Zürich
20. So KK, Chantrasmi T, Hu XY, Witteveen JAS, Stemmer C, Iaccarino G, Adams NA (2010) Uncertainty analysis for shock-bubble interaction. In: Proceedings of the 2010 summer program. Center for Turbulence Research, Stanford University
21. Sod GA (1978) A survey of several finite difference methods for systems of nonlinear hyperbolic conservation laws. J Comput Phys 27(1):1–31. doi:10.1016/0021-9991(78)90023-2, http://www.sciencedirect.com/science/article/pii/0021999178900232
22. Strand B (1994) Summation by parts for finite difference approximations for d/dx. J Comput Phys 110(1):47–67. doi:http://dx.doi.org/10.1006/jcph.1994.1005
23. van Leer B (1979) Towards the ultimate conservative difference scheme. V – a second-order sequel to Godunov's method. J Comput Phys 32:101–136. doi:10.1016/0021-9991(79)90145-1
24. Xiu D, Karniadakis GE (2002) The Wiener–Askey polynomial chaos for stochastic differential equations. SIAM J Sci Comput 24(2):619–644. doi:http://dx.doi.org/10.1137/S1064827501387826

Appendix A
Generation of Multiwavelets

Algorithm 1 Generation of multiwavelets (mother-wavelets (2.14))

Start with the set of functions $\{f_k^1\}_{k=0}^{N_p}$, defined by

$$f_k^1(\xi) = \begin{cases} \xi^k, & \xi \in [-1, 0], \\ -\xi^k, & \xi \in [0, 1], \\ 0, & \text{otherwise}. \end{cases}$$

STEP 1: Orthogonalize w.r.t. the monomials $1, \ldots, \xi^{N_p}$ (Gram-Schmidt) to obtain $\{f_k^2\}_{k=0}^{N_p}$.
STEP 2:
  for $i \leftarrow 0$ to $N_p - 1$ **do**
    Make sure $\langle f_i^{i+1} \xi^{N_p+i} \rangle \neq 0$ (otherwise reorder).
    for $j = i + 1$ to N_0 **do**
      $w = \frac{\langle f_j^{i+2} \xi^{N_p+i} \rangle}{\langle f_i^{i+2} \xi^{N_p+i} \rangle}$
      $f_j^{i+3} \leftarrow f_j^{i+2} - w f_i^{i+2}$
    end for
  end for
STEP 3: Orthogonalize $\{f_i^{i+2}\}_{i=0}^{N_p}$ using G-S.
  for $i \leftarrow N_p$ to 0 **do**
    $\psi_i^W(\xi) \leftarrow$ Apply Gram-Schmidt to f_i^{i+2}.
  end for
  Output $\{\psi_i^W(\xi)\}_{i=0}^{N_p}$.

M.P. Pettersson et al., *Polynomial Chaos Methods for Hyperbolic Partial Differential Equations*, Mathematical Engineering,
DOI 10.1007/978-3-319-10714-1

Appendix B
Proof of Constant Eigenvectors of Low-Order MW Triple Product Matrices

B.1 Proof of Constant Eigenvectors of A

Proposition B.1. *The matrix A defined by (8.3) for Haar wavelets $\{\psi_j\}_{j=0}^{M}$ has constant eigenvectors for all $M + 1 = 2^{N_r}$, $N_r \in \mathbb{N}$.*

Proof (Sketch of proof). We will use induction on the order M of wavelet chaos to show that the matrix A has constant eigenvectors for all orders M. In order to do this, we will need certain features of the structure of A. To facilitate the notation, denote $\tilde{M} = M + 1$. We can express $A_{2\tilde{M}}$ in terms of the matrix $A_{\tilde{M}}$. Two properties of the triple product $\langle \psi_i \psi_j \psi_k \rangle$ will be used to prove that A does indeed have the matrix structure presented.

Property 1. Let $i \in \{0, \dots, \tilde{M} - 1\}$, $j = k \in \{\tilde{M}, \dots, 2\tilde{M} - 1\}$ and let j' and j'' be the progenies of j. Then

$$\langle \psi_i \psi_j^2 \rangle = \langle \psi_i \psi_{j'}^2 \rangle = \langle \psi_i \psi_{j''}^2 \rangle.$$

Property 2. Consider the indices $i \in \{\tilde{M}, \dots, 2\tilde{M} - 1\}$, $j = k \in \{2\tilde{M}, \dots, 4\tilde{M} - 1\}$. Then

$$\langle \psi_i \psi_j^2 \rangle = \begin{cases} \tilde{M}^{1/2} & \text{if } j \text{ first progeny of } i \\ -\tilde{M}^{1/2} & \text{if } j \text{ second progeny of } i. \\ 0 & \text{otherwise} \end{cases}$$

M.P. Pettersson et al., *Polynomial Chaos Methods for Hyperbolic Partial Differential Equations*, Mathematical Engineering,
DOI 10.1007/978-3-319-10714-1

As an induction hypothesis, we assume that given $\boldsymbol{A}_{\tilde{M}}$ for some $\tilde{M} = 2^{N_r}$, $N_r \in \mathbb{N}$, the next order of triple product matrix $\boldsymbol{A}_{2\tilde{M}}$ can be written

$$\boldsymbol{A}_{2\tilde{M}} = \begin{bmatrix} \boldsymbol{A}_{\tilde{M}} & \boldsymbol{Q}_{\tilde{M}}\boldsymbol{D}_{\tilde{M}} \\ \boldsymbol{D}_{\tilde{M}}\boldsymbol{Q}_{\tilde{M}}^T & \boldsymbol{\Lambda} \end{bmatrix},$$

where $\boldsymbol{Q}_{\tilde{M}}$ is the matrix of constant eigenvectors of $\boldsymbol{A}_{\tilde{M}}$ satisfying $\|\boldsymbol{Q}_{\tilde{M}}\|_2^2 = \tilde{M}$, $\boldsymbol{D}_{\tilde{M}} = \mathrm{diag}(w_{\tilde{M}}, \ldots, w_{2\tilde{M}-1})$ and $\boldsymbol{\Lambda}$ is diagonal and contains the eigenvalues of $\boldsymbol{A}_{\tilde{M}}$. Then, we have that

$$\begin{bmatrix} \boldsymbol{A}_{\tilde{M}} & \boldsymbol{Q}_{\tilde{M}}\boldsymbol{D}_{\tilde{M}} \\ \boldsymbol{D}_{\tilde{M}}\boldsymbol{Q}_{\tilde{M}}^T & \boldsymbol{\Lambda} \end{bmatrix} \begin{bmatrix} \boldsymbol{Q}_{\tilde{M}} \\ \pm\tilde{M}^{1/2}\boldsymbol{I} \end{bmatrix} = \begin{bmatrix} \boldsymbol{Q}_{\tilde{M}}\boldsymbol{\Lambda} \pm \tilde{M}^{1/2}\boldsymbol{Q}\boldsymbol{D}_{\tilde{M}} \\ \tilde{M}\boldsymbol{D}_{\tilde{M}} \pm \tilde{M}^{1/2}\boldsymbol{\Lambda} \end{bmatrix} = \begin{bmatrix} \boldsymbol{Q}_{\tilde{M}} \\ \pm\tilde{M}^{1/2}\boldsymbol{I} \end{bmatrix} (\boldsymbol{\Lambda} \pm \tilde{M}^{1/2}\boldsymbol{D}_{\tilde{M}}),$$

so the eigenvalues and eigenvectors of $\boldsymbol{A}_{2\tilde{M}}$ are given by $\boldsymbol{\Lambda} \pm \tilde{M}^{1/2}\boldsymbol{D}_{\tilde{M}}$ and $[\boldsymbol{Q}_{\tilde{M}}, \pm\tilde{M}^{1/2}\boldsymbol{I}]^T$, respectively. For the next order of expansion, $4\tilde{M}$, we have

$$\boldsymbol{A}_{4\tilde{M}} = \begin{bmatrix} \begin{bmatrix} \boldsymbol{A}_{\tilde{M}} & \boldsymbol{Q}_{\tilde{M}}\boldsymbol{D}_{\tilde{M}} \\ \boldsymbol{D}_{\tilde{M}}\boldsymbol{Q}_{\tilde{M}}^T & \boldsymbol{\Lambda} \end{bmatrix} & \begin{bmatrix} \boldsymbol{Q}_{\tilde{M}} \otimes [1,1] \\ \tilde{M}^{1/2}\boldsymbol{I} \otimes [1,-1] \end{bmatrix} \boldsymbol{D}_{2\tilde{M}} \\ \boldsymbol{D}_{2\tilde{M}} \begin{bmatrix} \boldsymbol{Q}_{\tilde{M}} \otimes [1,1] \\ \tilde{M}^{1/2}\boldsymbol{I} \otimes [1,-1] \end{bmatrix}^T & \boldsymbol{\Lambda} \otimes \boldsymbol{I}_2 + \tilde{M}^{1/2}\boldsymbol{D}_{\tilde{M}} \otimes \begin{bmatrix} 1 & 0 \\ 0 & -1 \end{bmatrix} \end{bmatrix}. \tag{B.1}$$

To see that this is indeed the structure of $\boldsymbol{A}_{4\tilde{M}}$, note that any nonzero matrix entry not already present in $\boldsymbol{A}_{2\tilde{M}}$, can be deduced using properties 1 and 2, and scaling the rows/columns by multiplication by the diagonal matrix $\boldsymbol{D}_{2\tilde{M}}$. The structure of $\boldsymbol{A}_{4\tilde{M}}$ follows from the construction of the Haar wavelet basis, but we do not give a proof here.

One can verify that $\boldsymbol{A}_{4\tilde{M}}$ given by (B.1) has the eigenvectors and eigenvalues

$$\boldsymbol{Q}_{4\tilde{M}} = \begin{bmatrix} \begin{bmatrix} \boldsymbol{Q}_{\tilde{M}} \otimes [1,1] \\ \tilde{M}^{1/2}\boldsymbol{I}_{\tilde{M}} \otimes [1,-1] \end{bmatrix} \\ \pm(2\tilde{M})^{1/2}\boldsymbol{I}_{2\tilde{M}} \end{bmatrix},$$

$$\boldsymbol{\Lambda}_{4\tilde{M}} = \boldsymbol{\Lambda} \otimes \boldsymbol{I}_2 + \tilde{M}^{1/2}\boldsymbol{D}_{\tilde{M}} \otimes \begin{bmatrix} 1 & 0 \\ 0 & -1 \end{bmatrix} \pm (2\tilde{M})^{1/2}\boldsymbol{D}_{2\tilde{M}},$$

so the eigenvectors are constant (but the eigenvalues are variable in the coefficients $(w_i)_j$ through $M_{\tilde{M}}$ and $M_{2\tilde{M}}$). The base cases $\tilde{M} = 1$, $\tilde{M} = 2$, can easily be verified, so by induction $\boldsymbol{A}_{\tilde{M}}$ has constant eigenvectors for all $\tilde{M} = 2^{N_r}$, $N_r \in \mathbb{N}$.

B.2 Eigenvalue Decompositions of A

B.2.1 Piecewise Constant Multiwavelets (Haar Wavelets)

B.2.1.1 $N_r = 2$

$$Q = \frac{1}{4}\begin{bmatrix} 1 & 1 & 1 & 1 \\ 1 & 1 & -1 & -1 \\ \sqrt{2} & -\sqrt{2} & 0 & 0 \\ 0 & 0 & \sqrt{2} & -\sqrt{2} \end{bmatrix}, \quad \boldsymbol{\Lambda} = diag\begin{bmatrix} u_0 + u_1 + \sqrt{2}u_2 \\ u_0 + u_1 - \sqrt{2}u_2 \\ u_0 - u_1 + \sqrt{2}u_3 \\ u_0 - u_1 - \sqrt{2}u_3 \end{bmatrix}$$

B.2.1.2 $N_r = 3$

$$Q = \frac{1}{\sqrt{8}}\begin{bmatrix} 1 & 1 & 1 & 1 & 1 & 1 & 1 & 1 \\ 1 & 1 & 1 & 1 & -1 & -1 & -1 & -1 \\ \sqrt{2} & \sqrt{2} & -\sqrt{2} & -\sqrt{2} & 0 & 0 & 0 & 0 \\ 0 & 0 & 0 & 0 & \sqrt{2} & \sqrt{2} & -\sqrt{2} & -\sqrt{2} \\ 2 & -2 & 0 & 0 & 0 & 0 & 0 & 0 \\ 0 & 0 & 2 & -2 & 0 & 0 & 0 & 0 \\ 0 & 0 & 0 & 0 & 2 & -2 & 0 & 0 \\ 0 & 0 & 0 & 0 & 0 & 0 & 2 & -2 \end{bmatrix}$$

$$\boldsymbol{\Lambda} = diag\begin{bmatrix} u_0 + u_1 + \sqrt{2}u_2 + 2u_4 \\ u_0 + u_1 + \sqrt{2}u_2 - 2u_4 \\ u_0 + u_1 - \sqrt{2}u_2 + 2u_5 \\ u_0 + u_1 - \sqrt{2}u_2 - 2u_5 \\ u_0 - u_1 + \sqrt{2}u_3 + 2u_6 \\ u_0 - u_1 + \sqrt{2}u_3 - 2u_6 \\ u_0 - u_1 - \sqrt{2}u_3 + 2u_7 \\ u_0 - u_1 - \sqrt{2}u_3 - 2u_7 \end{bmatrix}$$

B.2.2 Piecewise Linear Multiwavelets

B.2.2.1 $N_r = 1$

$$Q = \begin{bmatrix} \frac{1}{2} & \frac{1}{2} & \frac{1}{2} & \frac{1}{2} \\ -\frac{\sqrt{3}+1}{4} & \frac{\sqrt{3}-1}{4} & \frac{\sqrt{3}+1}{4} & -\frac{\sqrt{3}-1}{4} \\ -\frac{1}{2} & \frac{1}{2} & -\frac{1}{2} & \frac{1}{2} \\ -\frac{\sqrt{3}-1}{4} & -\frac{\sqrt{3}+1}{4} & \frac{\sqrt{3}-1}{4} & -\frac{\sqrt{3}+1}{4} \end{bmatrix}, \quad \boldsymbol{\Lambda} = diag\begin{bmatrix} u_0 - \frac{\sqrt{3}+1}{2}u_1 - u_2 - \frac{\sqrt{3}-1}{2}u_3 \\ u_0 + \frac{\sqrt{3}-1}{2}u_1 + u_2 - \frac{\sqrt{3}+1}{2}u_3 \\ u_0 + \frac{\sqrt{3}+1}{2}u_1 - u_2 + \frac{\sqrt{3}-1}{2}u_3 \\ u_0 - \frac{\sqrt{3}-1}{2}u_1 + u_2 - \frac{\sqrt{3}+1}{2}u_3 \end{bmatrix}$$

B.2.2.2 $N_r = 2$

$$
\boldsymbol{Q}=\begin{bmatrix}
\frac{1}{\sqrt{8}} & \frac{1}{\sqrt{8}} & \frac{1}{\sqrt{8}} & \frac{1}{\sqrt{8}} & \frac{1}{\sqrt{8}} & \frac{1}{\sqrt{8}} & \frac{1}{\sqrt{8}} & \frac{1}{\sqrt{8}} \\
\frac{\sqrt{14+3\sqrt{3}}}{8} & -\frac{\sqrt{14+3\sqrt{3}}}{8} & -\frac{\sqrt{14-3\sqrt{3}}}{8} & \frac{\sqrt{14-3\sqrt{3}}}{8} & \frac{\sqrt{3}+1}{8\sqrt{2}} & -\frac{\sqrt{3}+1}{8\sqrt{2}} & -\frac{\sqrt{3}-1}{8\sqrt{2}} & \frac{\sqrt{3}-1}{8\sqrt{2}} \\
-\frac{\sqrt{3}+1}{4\sqrt{2}} & -\frac{\sqrt{3}+1}{4\sqrt{2}} & -\frac{\sqrt{3}-1}{4\sqrt{2}} & -\frac{\sqrt{3}-1}{4\sqrt{2}} & \frac{\sqrt{3}-1}{4\sqrt{2}} & \frac{\sqrt{3}-1}{4\sqrt{2}} & \frac{\sqrt{3}+1}{4\sqrt{2}} & \frac{\sqrt{3}+1}{4\sqrt{2}} \\
\frac{\sqrt{3}+1}{8\sqrt{2}} & -\frac{\sqrt{3}+1}{8\sqrt{2}} & \frac{\sqrt{3}-1}{8\sqrt{2}} & -\frac{\sqrt{3}-1}{8\sqrt{2}} & -\frac{\sqrt{14-5\sqrt{3}}}{8} & \frac{\sqrt{14-5\sqrt{3}}}{8} & \frac{\sqrt{14+5\sqrt{3}}}{8} & -\frac{\sqrt{14+5\sqrt{3}}}{8} \\
0 & -\frac{1}{2} & \frac{1}{2} & 0 & 0 & \frac{1}{2} & -\frac{1}{2} & 0 \\
0 & -\frac{\sqrt{3}-1}{4} & \frac{\sqrt{3}+1}{4} & 0 & 0 & -\frac{\sqrt{3}+1}{4} & \frac{\sqrt{3}-1}{4} & 0 \\
-\frac{1}{2} & 0 & 0 & \frac{1}{2} & \frac{1}{2} & 0 & 0 & -\frac{1}{2} \\
\frac{\sqrt{3}-1}{4} & 0 & 0 & -\frac{\sqrt{3}+1}{4} & \frac{\sqrt{3}+1}{4} & 0 & 0 & -\frac{\sqrt{3}-1}{4}
\end{bmatrix}
$$

$$
\boldsymbol{\Lambda} = diag\begin{bmatrix}
u_0 + \sqrt{\frac{14+3\sqrt{3}}{8}}u_1 - \frac{\sqrt{3}+1}{2}u_2 + \frac{\sqrt{3}+1}{4}u_3 - \sqrt{2}u_6 + \frac{\sqrt{3}-1}{\sqrt{2}}u_7 \\
u_0 - \sqrt{\frac{14+3\sqrt{3}}{8}}u_1 - \frac{\sqrt{3}+1}{2}u_2 - \frac{\sqrt{3}+1}{4}u_3 - \sqrt{2}u_4 - \frac{\sqrt{3}-1}{\sqrt{2}}u_5 \\
u_0 - \sqrt{\frac{14-3\sqrt{3}}{8}}u_1 - \frac{\sqrt{3}-1}{2}u_2 + \frac{\sqrt{3}-1}{4}u_3 + \sqrt{2}u_4 + \frac{\sqrt{3}+1}{\sqrt{2}}u_5 \\
u_0 + \sqrt{\frac{14-3\sqrt{3}}{8}}u_1 - \frac{\sqrt{3}-1}{2}u_2 - \frac{\sqrt{3}-1}{4}u_3 + \sqrt{2}u_6 - \frac{\sqrt{3}+1}{\sqrt{2}}u_7 \\
u_0 + \frac{\sqrt{3}+1}{4}u_1 + \frac{\sqrt{3}-1}{2}u_2 - \sqrt{\frac{14-5\sqrt{3}}{8}}u_3 + \sqrt{2}u_6 + \frac{\sqrt{3}+1}{\sqrt{2}}u_7 \\
u_0 - \frac{\sqrt{3}+1}{4}u_1 + \frac{\sqrt{3}-1}{2}u_2 + \sqrt{\frac{14-5\sqrt{3}}{8}}u_3 + \sqrt{2}u_4 - \frac{\sqrt{3}+1}{\sqrt{2}}u_5 \\
u_0 - \frac{\sqrt{3}-1}{4}u_1 + \frac{\sqrt{3}+1}{2}u_2 + \sqrt{\frac{14+5\sqrt{3}}{8}}u_3 - \sqrt{2}u_4 + \frac{\sqrt{3}-1}{\sqrt{2}}u_5 \\
u_0 + \frac{\sqrt{3}-1}{4}u_1 + \frac{\sqrt{3}+1}{2}u_2 - \sqrt{\frac{14+5\sqrt{3}}{8}}u_3 - \sqrt{2}u_6 - \frac{\sqrt{3}-1}{\sqrt{2}}u_7
\end{bmatrix}
$$

Appendix C
Matlab Codes

The reference codes used to generate the results presented in Chaps. 5 and 6 are reported here. They can also be downloaded from the website http://extras.springer.com/2015/

C.1 Linear Transport

C.1.1 Main Code

C.1.1.1 advection_diffusion_main.m

```
% Advection-diffusion with stochastic viscosity coefficient, SBP-SAT
% implementation
% Time integration with 4th order Runge-Kutta

clear all;

% Choose mod == 'herm' for Hermite polynomials representing mu w. lognormal
% distribution, or choose mod == 'lege' for Legendre polynomials and
% uniform distribution of mu.

mod = 'lege';

% Spatial interval end points
left = 0.4;
right = 0.6;

% Problem input parameters
v = 2;
M = 1;
rho_0 = 0.5;

c1= 0.02;
c2= 0.01;

t = 0;
t0 = 0.005;
x0 = 0.5;


p = 4; % Number of gPC coefficients
m = 50; % Number of discretization points in space
T = 0.001; % End time
```


M.P. Pettersson et al., *Polynomial Chaos Methods for Hyperbolic Partial Differential Equations*, Mathematical Engineering,
DOI 10.1007/978-3-319-10714-1

```
order = 6; % Order of accuracy of SBP operators, should be 2, 4 or 6.

% Setup parameters determined as functions of other parameters
dx = (right-left)/(m-1); % Spatial step length
dt = 0.2*dx^2/v; % Time step, here dependent on (Delta x)^2 but changes with ratio viscosity/velocity
V = v*eye(m*p); % Advective velocity, here assumed spatially uniform
x = linspace(left,right,m);


I_p=eye(p);
I_m=eye(m);

old = 1;
new = 2;

u = zeros(p*m,2);
%Compute inner triple products of the chosen basis functions
if mod == 'lege'
    C = Legendre_chaos(p-1);
end
if mod == 'herm'
    C = Hermite_chaos(p-1);
end

%Difference operators Dx=P^(-1)Q

[D1,D2,BD,D,Pnorm] = SBP_operators(m,dx,order);
P_inv = inv(Pnorm);
D1 = kron(D1,eye(p));
D2 = kron(D2,eye(p));


%Initialization and forcing terms
u_init = zeros(m*p,1);
if t0>0
    if mod == 'herm'
        u(:,old) = init_Hermite(rho_0,c1,c2,v,t,t0,x,x0,m,p);
    end
    if mod == 'lege'
        u(:,old) = init_Legendre(rho_0,c1,c2,v,t,t0,x,x0,m,p);
    end
end

% This is a first order approximation of the delta function
if t0 == 0
    u((m-1)/2*p+1,old) = rho_0/(dx);
end


if mod == 'herm'
    B1 = mu_SG_lognormal(c1,c2,p);
end
if mod == 'lege'
    if p>1
        B1 = C(:,:,1)*c1+C(:,:,2)*c2/sqrt(3);
    end
    if p==1
        B1 = C(:,:,1)*c1;
    end
end

% Compute force terms
E_0 = sparse(zeros(m));
E_0(1,1) = 1;
E_M = sparse(zeros(m));
E_M(m,m) = 1;
Sig_I_L = -v/2*eye(p);
Sig_V_L = B1;
Sig_V_R = -B1;
force_left_1 = sparse(kron(P_inv,I_p)*kron(E_0,Sig_I_L));
force_left_2 = kron(P_inv*D',I_p)*kron(E_0,Sig_V_L);
force_right = sparse(kron(P_inv,I_p)*kron(E_M,Sig_V_R));

g0 = zeros(m*p,1);
g1 = zeros(m*p,1);


B = kron(I_m,B1);

%Iterate over time with fourth order Runge-Kutta until time T is reached

while (t<T)
    if T-t<dt
        dt = T-t;
    end
    t = t+dt;
    if mod == 'herm'
        [g0 g1] = bdy_cond_Hermite(rho_0,c1,c2,v,t,t0,x,x0,m,p);
    end
```

```
    if mod == 'lege'
        [g0 g1] = bdy_cond_Legendre(rho_0,c1,c2,v,t,t0,x,x0,m,p);
    end

    % Fourth order Runge-Kutta in time
    k1 = dt*((-V*D1+B*D2)*u(:,old)+force_left_1*(u(:,old)-g0)+force_left_2*(u(:,old)-g0)+force_right*(kron(D,
        I_p)*u(:,old)-g1));
    k2 = dt*((-V*D1+B*D2)*(u(:,old)+k1/2)+force_left_1*(u(:,old)+k1/2-g0)+force_left_2*((u(:,old)+k1/2)-g0)+
        force_right*(kron(D,I_p)*(u(:,old)+k1/2)-g1));
    k3 = dt*((-V*D1+B*D2)*(u(:,old)+k2/2)+force_left_1*(u(:,old)+k2/2-g0)+force_left_2*((u(:,old)+k2/2)-g0)+
        force_right*(kron(D,I_p)*(u(:,old)+k2/2)-g1));
    k4 = dt*((-V*D1+B*D2)*(u(:,old)+k3)+force_left_1*(u(:,old)+k3-g0)+force_left_2*((u(:,old)+k3)-g0)+
        force_right*(kron(D,I_p)*(u(:,old)+k3)-g1));

    % Update the solution vector
    u(:,new)=u(:,old)+1/6*(k1+2*k2+2*k3+k4);
    u(:,old)=u(:,new);


    % Plot the solution
    for k=1:p
        u_plot(:,k) = u(k:p:end,new);
    end
    plot(x,u_plot,'-*');
    title(['gPC coefficients, t=' num2str(t,'%.4f')])
    leg_strs = {};
    for k=1:p
        legstrs{k} = ['gPC coe. ' num2str(k-1)];
    end

    legend(legstrs);
    drawnow;
end

u = u(:,new);

% Statistics are readily obtained from the gPC coefficients
u_num_exp = u(1:p:p*(m-1)+1);
u_num_var = zeros(m,1);

for i=1:m
    u_num_var(i,1) = sum(u((i-1)*p+2:i*p).^2);
end

% Compute reference statistics based on numerical quadrature of the exact
% solution
if mod == 'herm'
    [u_exp u_var] = statistics_adv_diff_lognorm(m,t,x,x0,c1,c2,rho_0,t0,v);
end
if mod == 'lege'
    [u_exp u_var] = statistics_adv_diff_uniform_mu(m,t,x,x0,c1,c2,rho_0,t0,v);
end

% Plot the computed and reference expectation and variance as a function of
% space (at time T)

figure;
plot(x,u_num_exp,'-r',x,u_exp,'--k','LineWidth',1.5);
h_legend = legend('Num','Ref');;
set(h_legend,'FontSize',12);
title('Mean','FontSize',16);

figure;
plot(x,u_num_var,'-r',x,u_var,'--k','LineWidth',1.5);
h_legend = legend('Num','Ref');;
set(h_legend,'FontSize',12);
title('Variance','FontSize',16);
```

C.1.1.2 Legendre_chaos.m

```

function [C] = legendre_chaos(n)

% Compute Legendre chaos parameters

% Indata:
% n — Order of gPC

% Outdata:
% C — Three term inner products C(i,j,k) = E[Phi_i Phi_j Phi_k]


C = zeros(n+1,n+1,n+1);
for i = 0:n
    for j = 0:n
        for k = 0:n
            s = (i+k+j) / 2;
            if rem(i+k+j,2) == 1 || abs(i-j) > k || k > i+j
                C(i+1,j+1,k+1) = 0;
            else
                C(i+1,j+1,k+1) = sqrt((2*i+1)*(2*j+1)*(2*k+1))/(i+j+k+1)*A_for_lege(s-i)*A_for_lege(s-j)*
        A_for_lege(s-k)/A_for_lege(s);
            end
        end
    end
end
return
```

C.1.1.3 Hermite_chaos.m

```

function [C] = Hermite_chaos(n)

% Compute hermite chaos parameters

% Indata:
% n — Order of gPC

% Outdata:
% C — Three term inner products C(i,j,k) = E[Phi_i Phi_j Phi_k]


C = zeros(n+1,n+1,n+1);
for i = 0:n
    for j = 0:n
        for k = 0:n
            s = (i+k+j) / 2;
            if rem(i+k+j,2) == 1 || i > s || j > s || k > s
                C(i+1,j+1,k+1) = 0;
            else
                C(i+1,j+1,k+1) = factorial(i) * factorial(j) ...
                    *factorial(k) / ((factorial(s-i) * factorial(s-j) * factorial(s-k)) * sqrt(factorial(i)) *
        sqrt(factorial(j)) * sqrt(factorial(k)));
            end
        end
    end
end
return
```

C.1.1.4 A_for_lege.m

```

function [A] = A_for_lege(n)

% Auxiliary function for triple product matrix of Legendre polynomials

if n==0
    A = 1;
end
if n < 0
    A = 0;
end
if n >= 1
    A = 1;
    for j=1:n
        A = A*(2*j-1);
    end
    A = A/factorial(n);
end
```

C.1.2 Discretization Operators

C.1.2.1 SBP_operators.m

```
function [D1 D2,BS,S,H] = SBP_operators(n,dx,order)

% SBP operators of orders 2, 4, 6 and 8 for the first and second derivative.

% Indata:
% n — Number of spatial grid pts
% dx — Step size
% order — Order of accuracy (only for 2,4,6,8)

% Outdata:
% D1 — First derivative operator
% D2 — Second derivative operator (D = P^{—1}M)
% S — First derivative operator on boundaries
% BS — The boundary elements in the energy estimate
% H — The norm operator (denoted P in some papers)


e = ones(n,1);

if order==2

    D1 = 1/dx*spdiags([−1/2*e 0*e 1/2*e],−1:1,n,n);
    D1(1,1) = −1/dx;
    D1(1,2) = 1/dx;
    D1(1,3) = 0;
    D1(n,n) = 1/dx;
    D1(n,n−1) = −1/dx;
    D1(n,n−2) = 0;

    %%%%%%
    D2 = 1/(dx^2)*spdiags([1*e −2*e 1*e],−1:1,n,n);
    D2(1,1) = 1/(dx^2);
    D2(1,2) = −2/(dx^2);
    D2(1,3) = 1/(dx^2);
    D2(n,n) = 1/(dx^2);
    D2(n,n−1) = −2/(dx^2);
    D2(n,n−2) = 1/(dx^2);


    H = dx*spdiags([e],0,n,n);
    H(1,1) = dx*1/2;
    H(n,n) = dx*1/2;

    BS = (1/dx)*spdiags(zeros(size(e)),0,n,n);
    BS(1,1) = 3/2/dx;
    BS(1,2) = −2/dx;
    BS(1,3) = 1/2/dx;
    BS(n,n) = 3/2/dx;
    BS(n,n−1) = −2/dx;
    BS(n,n−2) = 1/2/dx;

    S = (1/dx)*spdiags([e],0,n,n);
    S(1,1) = −3/2/dx;
    S(1,2) = 2/dx;
    S(1,3) = −1/2/dx;

    S(n,n) = 3/2/dx;
    S(n,n−1) = −2/dx;
    S(n,n−2) = 1/2/dx;

elseif order==4

    D1 = 1/dx*spdiags([1/12*e −2/3*e 0*e 2/3*e −1/12*e],−2:2,n,n);
    D1(1,1) = −24/17/dx;
    D1(1,2) = 59/34/dx;
    D1(1,3) = −4/17/dx;
    D1(1,4) = −3/34/dx;
    D1(1,5) = 0;
    D1(1,6) = 0;
    D1(2,1) = −1/2/dx;
    D1(2,2) = 0;
    D1(2,3) = 1/2/dx;
    D1(2,4:6) = 0;
    D1(3,1) = 4/43/dx;
    D1(3,2) = −59/86/dx;
    D1(3,3) = 0;
    D1(3,4) = 59/86/dx;
    D1(3,5) = −4/43/dx;
    D1(3,6) = 0;
    D1(4,1) = 3/98/dx;
    D1(4,2) = 0;
```

```
D1(4,3) = -59/98/dx;
D1(4,4) = 0;
D1(4,5) = 32/49/dx;
D1(4,6) = -4/49/dx;
D1(4,7) = 0;
D1(n,n) = -D1(1,1);
D1(n,n-1) = -D1(1,2);
D1(n,n-2) = -D1(1,3);
D1(n,n-3) = -D1(1,4);
D1(n,n-4) = -D1(1,5);
D1(n,n-5) = -D1(1,6);
D1(n-1,n) = -D1(2,1);
D1(n-1,n-1) = -D1(2,2);
D1(n-1,n-2) = -D1(2,3);
D1(n-1,n-3) = -D1(2,4);
D1(n-1,n-4) = -D1(2,5);
D1(n-1,n-5) = -D1(2,6);
D1(n-2,n) = -D1(3,1);
D1(n-2,n-1) = -D1(3,2);
D1(n-2,n-2) = -D1(3,3);
D1(n-2,n-3) = -D1(3,4);
D1(n-2,n-4) = -D1(3,5);
D1(n-2,n-5) = -D1(3,6);
D1(n-3,n) = -D1(4,1);
D1(n-3,n-1) = -D1(4,2);
D1(n-3,n-2) = -D1(4,3);
D1(n-3,n-3) = -D1(4,4);
D1(n-3,n-4) = -D1(4,5);
D1(n-3,n-5) = -D1(4,6);


%%%%%%
D2 = 1/(dx^2)*spdiags([-1/12*e 4/3*e -5/2*e 4/3*e -1/12*e],-2:2,n,n);
D2(1,1) = 2/(dx^2);
D2(1,2) = -5/(dx^2);
D2(1,3) = 4/(dx^2);
D2(1,4) = -1/(dx^2);
D2(2,1) = 1/(dx^2);
D2(2,2) = -2/(dx^2);
D2(2,3) = 1/(dx^2);
D2(2,4) = 0;
D2(3,1) = -4/43/(dx^2);
D2(3,2) = 59/43/(dx^2);
D2(3,3) = -110/43/(dx^2);
D2(3,4) = 59/43/(dx^2);
D2(3,5) = -4/43/(dx^2);
D2(4,1) = -1/49/(dx^2);
D2(4,2) = 0;
D2(4,3) = 59/49/(dx^2);
D2(4,4) = -118/49/(dx^2);
D2(4,5) = 64/49/(dx^2);
D2(4,6) = -4/49/(dx^2);
D2(n,n) = D2(1,1);
D2(n,n-1) = D2(1,2);
D2(n,n-2) = D2(1,3);
D2(n,n-3) = D2(1,4);
D2(n-1,n) = D2(2,1);
D2(n-1,n-1) = D2(2,2);
D2(n-1,n-2) = D2(2,3);
D2(n-1,n-3) = D2(2,4);
D2(n-2,n) = D2(3,1);
D2(n-2,n-1) = D2(3,2);
D2(n-2,n-2) = D2(3,3);
D2(n-2,n-3) = D2(3,4);
D2(n-2,n-4) = D2(3,5);
D2(n-3,n) = D2(4,1);
D2(n-3,n-1) = D2(4,2);
D2(n-3,n-2) = D2(4,3);
D2(n-3,n-3) = D2(4,4);
D2(n-3,n-4) = D2(4,5);
D2(n-3,n-5) = D2(4,6);

H = dx*spdiags(e,0,n,n);
H(1,1) = dx*17/48;
H(2,2) = dx*59/48;
H(3,3) = dx*43/48;
H(4,4) = dx*49/48;
H(n,n) = H(1,1);
H(n-1,n-1) = H(2,2);
H(n-2,n-2) = H(3,3);
H(n-3,n-3) = H(4,4);

S = (1/dx)*spdiags(e,0,n,n);
S(1,1) = -11/6/dx;
S(1,2) = 3/dx;
S(1,3) = -3/2/dx;
S(1,4) = 1/3/dx;
S(n,n) = 11/6/dx;
S(n,n-1) = -3/dx;
```

```
    S(n,n-2) = 3/2/dx;
    S(n,n-3) = -1/3/dx;

    BS = (1/dx)*spdiags(zeros(size(e)),0,n,n);
    BS(1,1) = 11/6/dx;
    BS(1,2) = -3/dx;
    BS(1,3) = 3/2/dx;
    BS(1,4) = -1/3/dx;
    BS(n,n) = 11/6/dx;
    BS(n,n-1) = -3/dx;
    BS(n,n-2) = 3/2/dx;
    BS(n,n-3) = -1/3/dx;

elseif order==6
    e = ones(n,1);
    %%%%
    D1 = (1/(dx))*spdiags([-1/60*e 3/20*e -3/4*e 0*e 3/4*e -3/20*e 1/60*e],-3:3,n,n);

    D1(1,1) = -21600/13649/dx;
    D1(1,2) = 104009/54596/dx;
    D1(1,3) = 30443/81894/dx;
    D1(1,4) = -33311/27298/dx;
    D1(1,5) = 16863/27298/dx;
    D1(1,6) = -15025/163788/dx;
    D1(1,7) = 0;
    D1(1,8) = 0;
    D1(2,1) = -104009/240260/dx;
    D1(2,2) = 0;
    D1(2,3) = -311/72078/dx;
    D1(2,4) = 20229/24026/dx;
    D1(2,5) = -24337/48052/dx;
    D1(2,6) = 36661/360390/dx;
    D1(2,7) = 0;
    D1(2,8) = 0;
    D1(3,1) = -30443/162660/dx;
    D1(3,2) = 311/32532/dx;
    D1(3,3) = 0;
    D1(3,4) = -11155/16266/dx;
    D1(3,5) = 41287/32532/dx;
    D1(3,6) = -21999/54220/dx;
    D1(3,7) = 0;
    D1(3,8) = 0;
    D1(4,1) = 33311/107180/dx;
    D1(4,2) = -20229/21436/dx;
    D1(4,3) = 485/1398/dx;
    D1(4,4) = 0;
    D1(4,5) = 4147/21436/dx;
    D1(4,6) = 25427/321540/dx;
    D1(4,7) = 72/5359/dx;
    D1(4,8) = 0;
    D1(5,1) = -16863/78770/dx;
    D1(5,2) = 24337/31508/dx;
    D1(5,3) = -41287/47262/dx;
    D1(5,4) = -4147/15754/dx;
    D1(5,5) = 0;
    D1(5,6) = 342523/472620/dx;
    D1(5,7) = -1296/7877/dx;
    D1(5,8) = 144/7877/dx;
    D1(5,9) = 0;
    D1(6,1) = 15025/525612/dx;
    D1(6,2) = -36661/262806/dx;
    D1(6,3) = 21999/87602/dx;
    D1(6,4) = -25427/262806/dx;
    D1(6,5) = -342523/525612/dx;
    D1(6,6) = 0;
    D1(6,7) = 32400/43801/dx;
    D1(6,8) = -6480/43801/dx;
    D1(6,9) = 720/43801/dx;
    D1(6,10) = 0;
    D1(n,n) = -D1(1,1);
    D1(n,n-1) = -D1(1,2);
    D1(n,n-2) = -D1(1,3);
    D1(n,n-3) = -D1(1,4);
    D1(n,n-4) = -D1(1,5);
    D1(n,n-5) = -D1(1,6);
    D1(n,n-6) = -D1(1,7);
    D1(n,n-7) = -D1(1,8);
    D1(n-1,n) = -D1(2,1);
    D1(n-1,n-1) = -D1(2,2);
    D1(n-1,n-2) = -D1(2,3);
    D1(n-1,n-3) = -D1(2,4);
    D1(n-1,n-4) = -D1(2,5);
    D1(n-1,n-5) = -D1(2,6);
    D1(n-1,n-6) = -D1(2,7);
    D1(n-1,n-7) = -D1(2,8);
    D1(n-2,n) = -D1(3,1);
    D1(n-2,n-1) = -D1(3,2);
    D1(n-2,n-2) = -D1(3,3);
    D1(n-2,n-3) = -D1(3,4);
```

```
D1(n-2,n-4) = -D1(3,5);
D1(n-2,n-5) = -D1(3,6);
D1(n-2,n-6) = -D1(3,7);
D1(n-2,n-7) = -D1(3,8);
D1(n-3,n) = -D1(4,1);
D1(n-3,n-1) = -D1(4,2);
D1(n-3,n-2) = -D1(4,3);
D1(n-3,n-3) = -D1(4,4);
D1(n-3,n-4) = -D1(4,5);
D1(n-3,n-5) = -D1(4,6);
D1(n-3,n-6) = -D1(4,7);
D1(n-3,n-7) = -D1(4,8);
D1(n-4,n) = -D1(5,1);
D1(n-4,n-1) = -D1(5,2);
D1(n-4,n-2) = -D1(5,3);
D1(n-4,n-3) = -D1(5,4);
D1(n-4,n-4) = -D1(5,5);
D1(n-4,n-5) = -D1(5,6);
D1(n-4,n-6) = -D1(5,7);
D1(n-4,n-7) = -D1(5,8);
D1(n-4,n-8) = -D1(5,9);
D1(n-5,n) = -D1(6,1);
D1(n-5,n-1) = -D1(6,2);
D1(n-5,n-2) = -D1(6,3);
D1(n-5,n-3) = -D1(6,4);
D1(n-5,n-4) = -D1(6,5);
D1(n-5,n-5) = -D1(6,6);
D1(n-5,n-6) = -D1(6,7);
D1(n-5,n-7) = -D1(6,8);
D1(n-5,n-8) = -D1(6,9);
D1(n-5,n-9) = -D1(6,10);

%%%%
D2 = (1/(dx^2))*spdiags([1/90*e -3/20*e 3/2*e -49/18*e 3/2*e -3/20*e 1/90*e],-3:3,n,n);

D2(1,1) = 114170/40947/(dx^2);
D2(1,2) = -438107/54596/(dx^2);
D2(1,3) = 336409/40947/(dx^2);
D2(1,4) = -276997/81894/(dx^2);
D2(1,5) = 3747/13649/(dx^2);
D2(1,6) = 21035/163788/(dx^2);
D2(1,7) = 0;
D2(1,8) = 0;
D2(2,1) = 6173/5860/(dx^2);
D2(2,2) = -2066/879/(dx^2);
D2(2,3) = 3283/1758/(dx^2);
D2(2,4) =-303/293/(dx^2);
D2(2,5) = 2111/3516/(dx^2);
D2(2,6) = -601/4395/(dx^2);
D2(2,7) = 0;
D2(2,8) = 0;
D2(3,1) = -52391/81330/(dx^2);
D2(3,2) = 134603/32532/(dx^2);
D2(3,3) = -21982/2711/(dx^2);
D2(3,4) = 112915/16266/(dx^2);
D2(3,5) = -46969/16266/(dx^2);
D2(3,6) = 30409/54220/(dx^2);
D2(3,7) = 0;
D2(3,8) = 0;
D2(4,1) = 68603/321540/(dx^2);
D2(4,2) = -12423/10718/(dx^2);
D2(4,3) = 112915/32154/(dx^2);
D2(4,4) = -75934/16077/(dx^2);
D2(4,5) = 53369/21436/(dx^2);
D2(4,6) = -54899/160770/(dx^2);
D2(4,7) = 48/5359/(dx^2);
D2(4,8) = 0;
D2(5,1) = -7053/39385/(dx^2);
D2(5,2) = 86551/94524/(dx^2);
D2(5,3) = -46969/23631/(dx^2);
D2(5,4) = 53369/15754/(dx^2);
D2(5,5) = -87904/23631/(dx^2);
D2(5,6) = 820271/472620/( dx^2);
D2(5,7) = -1296/7877/(dx^2);
D2(5,8) = 96/7877/(dx^2);
D2(5,9) = 0;
D2(6,1) = 21035/525612/(dx^2);
D2(6,2) = -24641/131403/(dx^2);
D2(6,3) = 30409/87602/(dx^2);
D2(6,4) = -54899/131403/(dx^2);
D2(6,5) = 820271/525612/( dx^2);
D2(6,6) = -117600/43801/(dx^2);
D2(6,7) = 64800/43801/(dx^2);
D2(6,8) = -6480/43801/(dx^2);
D2(6,9) = 480/43801/(dx^2);
D2(6,10) = 0;
D2(n,n) = D2(1,1);
D2(n,n-1) = D2(1,2);
D2(n,n-2) = D2(1,3);
```

```
D2(n,n-3) = D2(1,4);
D2(n,n-4) = D2(1,5);
D2(n,n-5) = D2(1,6);
D2(n,n-6) = D2(1,7);
D2(n,n-7) = D2(1,8);
D2(n-1,n) = D2(2,1);
D2(n-1,n-1) = D2(2,2);
D2(n-1,n-2) = D2(2,3);
D2(n-1,n-3) = D2(2,4);
D2(n-1,n-4) = D2(2,5);
D2(n-1,n-5) = D2(2,6);
D2(n-1,n-6) = D2(2,7);
D2(n-1,n-7) = D2(2,8);
D2(n-2,n) = D2(3,1);
D2(n-2,n-1) = D2(3,2);
D2(n-2,n-2) = D2(3,3);
D2(n-2,n-3) = D2(3,4);
D2(n-2,n-4) = D2(3,5);
D2(n-2,n-5) = D2(3,6);
D2(n-2,n-6) = D2(3,7);
D2(n-2,n-7) = D2(3,8);
D2(n-3,n) = D2(4,1);
D2(n-3,n-1) = D2(4,2);
D2(n-3,n-2) = D2(4,3);
D2(n-3,n-3) = D2(4,4);
D2(n-3,n-4) = D2(4,5);
D2(n-3,n-5) = D2(4,6);
D2(n-3,n-6) = D2(4,7);
D2(n-3,n-7) = D2(4,8);
D2(n-4,n) = D2(5,1);
D2(n-4,n-1) = D2(5,2);
D2(n-4,n-2) = D2(5,3);
D2(n-4,n-3) = D2(5,4);
D2(n-4,n-4) = D2(5,5);
D2(n-4,n-5) = D2(5,6);
D2(n-4,n-6) = D2(5,7);
D2(n-4,n-7) = D2(5,8);
D2(n-4,n-8) = D2(5,9);
D2(n-5,n) = D2(6,1);
D2(n-5,n-1) = D2(6,2);
D2(n-5,n-2) = D2(6,3);
D2(n-5,n-3) = D2(6,4);
D2(n-5,n-4) = D2(6,5);
D2(n-5,n-5) = D2(6,6);
D2(n-5,n-6) = D2(6,7);
D2(n-5,n-7) = D2(6,8);
D2(n-5,n-8) = D2(6,9);
D2(n-5,n-9) = D2(6,10);

H = dx*spdiags([e],0,n,n);

H(1,1) = dx*13649/43200;
H(2,2) = dx*12013/8640;
H(3,3) = dx*2711/4320;
H(4,4) = dx*5359/4320;
H(5,5) = dx*7877/8640;
H(6,6) = dx*43801/43200;
H(n,n) = H(1,1);
H(n-1,n-1) = H(2,2);
H(n-2,n-2) = H(3,3);
H(n-3,n-3) = H(4,4);
H(n-4,n-4) = H(5,5);
H(n-5,n-5) = H(6,6);

BS = (1/dx)*spdiags([zeros(size(e))],0,n,n);

BS(1,1) = 25/12/dx;
BS(1,2) = -4/dx;
BS(1,3) = 3/dx;
BS(1,4) = -4/3/dx;
BS(1,5) = 1/4/dx;
BS(n,n) = BS(1,1);
BS(n,n-1) = BS(1,2);
BS(n,n-2) = BS(1,3);
BS(n,n-3) = BS(1,4);
BS(n,n-4) = BS(1,5);

S = (1/dx)*spdiags([e],0,n,n);

S(1,1) = -25/12/dx;
S(1,2) = 4/dx;
S(1,3) = -3/dx;
S(1,4) = 4/3/dx;
S(1,5) = -1/4/dx;
S(n,n) = BS(1,1);
S(n,n-1) = BS(1,2);
S(n,n-2) = BS(1,3);
S(n,n-3) = BS(1,4);
S(n,n-4) = BS(1,5);
```

```
elseif order==8
    e = ones(n,1);

    D = (1/(dx^2))*spdiags([-1/560*e 8/315*e -1/5*e 8/5*e -205/72*e 8/5*e -1/5*e 8/315*e -1/560*e],-4:4,n,n);
    % eighth order standard central stencil

    D(1,1) = 4870382994799/1358976868290/(dx^2);
    D(1,2) = -893640087518/75498714905/(dx^2);
    D(1,3) = 926594825119/60398971924/(dx^2);
    D(1,4) = -1315109406200/135897686829/(dx^2);
    D(1,5) = 39126983272/15099742981/(dx^2);
    D(1,6) = 12344491342/75498714905/(dx^2);
    D(1,7) = -451560522577/2717953736580/(dx^2);
    D(1,8) = 0;
    D(1,9) = 0;
    D(1,10) = 0;
    D(1,11) = 0;
    D(1,12) = 0;
    D(2,1) = 333806012194/390619153855/(dx^2);
    D(2,2) = -154646272029/111605472530/(dx^2);
    D(2,3) = 1168338040/33481641759/(dx^2);
    D(2,4) = 82699112501/133926567036/(dx^2);
    D(2,5) = -171562838/11160547253/(dx^2);
    D(2,6) = -28244698346/167408208795/(dx^2);
    D(2,7) = 11904122576/167408208795/(dx^2);
    D(2,8) = -2598164715/312495323084/(dx^2);
    D(2,9) = 0;
    D(2,10) = 0;
    D(2,11) = 0;
    D(2,12) = 0;
    D(3,1) = 7838984095/52731029988/(dx^2);
    D(3,2) = 1168338040/5649753213/(dx^2);
    D(3,3) = -88747895/144865467/(dx^2);
    D(3,4) = 423587231/627750357/(dx^2);
    D(3,5) = -43205598281/22599012852/(dx^2);
    D(3,6) = 4876378562/1883251071/(dx^2);
    D(3,7) = -5124426509/3766502142/(dx^2);
    D(3,8) = 10496900965/39548272491/(dx^2);
    D(3,9) = 0;
    D(3,10) = 0;
    D(3,11) = 0;
    D(3,12) = 0;
    D(4,1) = -94978241528/828644350023/(dx^2);
    D(4,2) = 82699112501/157837019052/(dx^2);
    D(4,3) = 1270761693/13153084921/(dx^2);
    D(4,4) = -167389605005/118377764289/(dx^2);
    D(4,5) = 48242560214/39459254763/(dx^2);
    D(4,6) = -31673996013/52612339684/(dx^2);
    D(4,7) = 43556319241/118377764289/(dx^2);
    D(4,8) = -44430275135/552429566682/(dx^2);
    D(4,9) = 0;
    D(4,10) = 0;
    D(4,11) = 0;
    D(4,12) = 0;
    D(5,1) = 1455067816/21132528431/(dx^2);
    D(5,2) = -171562838/3018932633/(dx^2);
    D(5,3) = -43205598281/36227191596/(dx^2);
    D(5,4) = 48242560214/9056797899/(dx^2);
    D(5,5) = -52276055645/6037865266/(dx^2);
    D(5,6) = 57521587238/9056797899/(dx^2);
    D(5,7) = -80321706377/36227191596/(dx^2);
    D(5,8) = 8078087158/21132528431/(dx^2);
    D(5,9) = -1296/299527/(dx^2);
    D(5,10) = 0;
    D(5,11) = 0;
    D(5,12) = 0;
    D(6,1) = 10881504334/327321118845/(dx^2);
    D(6,2) = -28244698346/140280479505/(dx^2);
    D(6,3) = 4876378562/9352031967/(dx^2);
    D(6,4) = -10557998671/12469375956/(dx^2);
    D(6,5) = 57521587238/28056095901/(dx^2);
    D(6,6) = -278531401019/93520319670/(dx^2);
    D(6,7) = 73790130002/46760159835/(dx^2);
    D(6,8) = -137529995233/785570685228/(dx^2);
    D(6,9) = 2048/103097/(dx^2);
    D(6,10) = -144/103097/(dx^2);
    D(6,11) = 0;
    D(6,12) = 0;
    D(7,1) = -135555328849/8509847458140/(dx^2);
    D(7,2) = 11904122576/101307707835/(dx^2);
    D(7,3) = -5124426509/13507694378/(dx^2);
    D(7,4) = 43556319241/60784624701/(dx^2);
    D(7,5) = -80321706377/81046166268/(dx^2);
    D(7,6) = 73790130002/33769235945/(dx^2);
    D(7,7) = -950494905688/303923123505/(dx^2);
    D(7,8) = 239073018673/141830790969/(dx^2);
    D(7,9) = -145152/670091/(dx^2);
    D(7,10) = 18432/670091/(dx^2);
    D(7,11) = -1296/670091/(dx^2);
```

```
D(7,12) = 0;
D(8,1) = 0;
D(8,2) = -2598164715/206729925524/(dx^2);
D(8,3) = 10496900965/155047444143/(dx^2);
D(8,4) = -44430275135/310094888286/(dx^2);
D(8,5) = 425162482/2720130599/(dx^2);
D(8,6) = -137529995233/620189776572/(dx^2);
D(8,7) = 239073018673/155047444143/(dx^2);
D(8,8) = -144648000000/51682481381/(dx^2);
D(8,9) = 8128512/5127739/(dx^2);
D(8,10) = -1016064/5127739/(dx^2);
D(8,11) = 129024/5127739/(dx^2);
D(8,12) = -9072/5127739/(dx^2);

D(n,n) = D(1,1);
D(n,n-1) = D(1,2);
D(n,n-2) = D(1,3);
D(n,n-3) = D(1,4);
D(n,n-4) = D(1,5);
D(n,n-5) = D(1,6);
D(n,n-6) = D(1,7);
D(n,n-7) = D(1,8);
D(n,n-8) = D(1,9);
D(n,n-9) = D(1,10);
D(n,n-10) = D(1,11);
D(n,n-11) = D(1,12);
D(n-1,n) = D(2,1);
D(n-1,n-1) = D(2,2);
D(n-1,n-2) = D(2,3);
D(n-1,n-3) = D(2,4);
D(n-1,n-4) = D(2,5);
D(n-1,n-5) = D(2,6);
D(n-1,n-6) = D(2,7);
D(n-1,n-7) = D(2,8);
D(n-1,n-8) = D(2,9);
D(n-1,n-9) = D(2,10);
D(n-1,n-10) = D(2,11);
D(n-1,n-11) = D(2,12);
D(n-2,n) = D(3,1);
D(n-2,n-1) = D(3,2);
D(n-2,n-2) = D(3,3);
D(n-2,n-3) = D(3,4);
D(n-2,n-4) = D(3,5);
D(n-2,n-5) = D(3,6);
D(n-2,n-6) = D(3,7);
D(n-2,n-7) = D(3,8);
D(n-2,n-8) = D(3,9);
D(n-2,n-9) = D(3,10);
D(n-2,n-10) = D(3,11);
D(n-2,n-11) = D(3,12);
D(n-3,n) = D(4,1);
D(n-3,n-1) = D(4,2);
D(n-3,n-2) = D(4,3);
D(n-3,n-3) = D(4,4);
D(n-3,n-4) = D(4,5);
D(n-3,n-5) = D(4,6);
D(n-3,n-6) = D(4,7);
D(n-3,n-7) = D(4,8);
D(n-3,n-8) = D(4,9);
D(n-3,n-9) = D(4,10);
D(n-3,n-10) = D(4,11);
D(n-3,n-11) = D(4,12);
D(n-4,n) = D(5,1);
D(n-4,n-1) = D(5,2);
D(n-4,n-2) = D(5,3);
D(n-4,n-3) = D(5,4);
D(n-4,n-4) = D(5,5);
D(n-4,n-5) = D(5,6);
D(n-4,n-6) = D(5,7);
D(n-4,n-7) = D(5,8);
D(n-4,n-8) = D(5,9);
D(n-4,n-9) = D(5,10);
D(n-4,n-10) = D(5,11);
D(n-4,n-11) = D(5,12);
D(n-5,n) = D(6,1);
D(n-5,n-1) = D(6,2);
D(n-5,n-2) = D(6,3);
D(n-5,n-3) = D(6,4);
D(n-5,n-4) = D(6,5);
D(n-5,n-5) = D(6,6);
D(n-5,n-6) = D(6,7);
D(n-5,n-7) = D(6,8);
D(n-5,n-8) = D(6,9);
D(n-5,n-9) = D(6,10);
D(n-5,n-10) = D(6,11);
D(n-5,n-11) = D(6,12);
D(n-6,n) = D(7,1);
D(n-6,n-1) = D(7,2);
D(n-6,n-2) = D(7,3);
```

```
    D(n-6,n-3) = D(7,4);
    D(n-6,n-4) = D(7,5);
    D(n-6,n-5) = D(7,6);
    D(n-6,n-6) = D(7,7);
    D(n-6,n-7) = D(7,8);
    D(n-6,n-8) = D(7,9);
    D(n-6,n-9) = D(7,10);
    D(n-6,n-10) = D(7,11);
    D(n-6,n-11) = D(7,12);
    D(n-7,n) = D(8,1);
    D(n-7,n-1) = D(8,2);
    D(n-7,n-2) = D(8,3);
    D(n-7,n-3) = D(8,4);
    D(n-7,n-4) = D(8,5);
    D(n-7,n-5) = D(8,6);
    D(n-7,n-6) = D(8,7);
    D(n-7,n-7) = D(8,8);
    D(n-7,n-8) = D(8,9);
    D(n-7,n-9) = D(8,10);
    D(n-7,n-10) = D(8,11);
    D(n-7,n-11) = D(8,12);


    H = dx*spdiags([e],0,n,n);

    H(1,1) = dx*1498139/5080320;
    H(2,2) = dx*1107307/725760;
    H(3,3) = dx*20761/80640;
    H(4,4) = dx*1304999/725760;
    H(5,5) = dx*299527/725760;
    H(6,6) = dx*103097/80640;
    H(7,7) = dx*670091/725760;
    H(8,8) = dx*5127739/5080320;
    H(n,n) = H(1,1);
    H(n-1,n-1) = H(2,2);
    H(n-2,n-2) = H(3,3);
    H(n-3,n-3) = H(4,4);
    H(n-4,n-4) = H(5,5);
    H(n-5,n-5) = H(6,6);
    H(n-6,n-6) = H(7,7);
    H(n-7,n-7) = H(8,8);

    BS = (1/dx)*spdiags([zeros(size(e))],0,n,n);

    BS(1,1) = 4723/2100/dx;
    BS(1,2) = -839/175/dx;
    BS(1,3) = 157/35/dx;
    BS(1,4) = -278/105/dx;
    BS(1,5) = 103/140/dx;
    BS(1,6) = 1/175/dx;
    BS(1,7) = -6/175/dx;
    BS(n,n) = BS(1,1);
    BS(n,n-1) = BS(1,2);
    BS(n,n-2) = BS(1,3);
    BS(n,n-3) = BS(1,4);
    BS(n,n-4) = BS(1,5);
    BS(n,n-5) = BS(1,6);
    BS(n,n-6) = BS(1,7);


    S = (1/dx)*spdiags([e],0,n,n);

    S(1,1) = -4723/2100/dx;
    S(1,2) = 839/175/dx;
    S(1,3) = -157/35/dx;
    S(1,4) = 278/105/dx;
    S(1,5) = -103/140/dx;
    S(1,6) = -1/175/dx;
    S(1,7) = 6/175/dx;
    S(n,n) = BS(1,1);
    S(n,n-1) = BS(1,2);
    S(n,n-2) = BS(1,3);
    S(n,n-3) = BS(1,4);
    S(n,n-4) = BS(1,5);
    S(n,n-5) = BS(1,6);
    S(n,n-6) = BS(1,7);

else
    disp('Only order 2, 4, 6 or 8 implemented here.')
end
```

C.1.2.2 mu_SG_lognormal.m

```
function [B] = mu_SG_lognormal(c1,c2,P)

% Compute the stochastic Galerkin viscosity matrix B with normalized Hermite
% polynomials

% Indata:
% c1, c2 — Scaling parameters of shifted lognormal distribution
% P — Number of gPC terms to be retained

% Outdata:
% B — Viscosity matrix, [B]_{ij} = sum_{k}^{2P}<psi_i psi_j psi_k> mu_k

% For the 1D case, use twice as many basis functions as for the variables
% (see Proposition 1 in Chapter 5)

P2 = 2*P;
C = hermite_chaos(P2-1);

tol = 30;

% Recursively generate Hermite polynomials
basis_fun = cell(1,P2);
basis_fun{1} = @(xi) xi.^0;
basis_fun{2} = @(xi) xi;
for k=3:P2
    basis_fun{k} = @(xi) 1/sqrt(k-1)*xi.*basis_fun{k-1}(xi)-sqrt((k-2)/(k-1))*basis_fun{k-2}(xi);
end

visc_fun = @(x) c1+c2*exp(x);

B = zeros(P2);

for k=1:P2
    integ = @(x) 1/sqrt(2*pi).*exp(-x.^2/2).*basis_fun{k}(x).*visc_fun(x);
    visc_hc(k,1) = quadgk(integ,-tol,tol);
    B = B+C(:,:,k)*visc_hc(k,1);
end

B = B(1:P,1:P);
```

C.1.3 Boundary Treatment

C.1.3.1 bdy_cond_Dirichlet.m

```
function [g0] = bdy_cond_Dirichlet(rho_0,c1,c2,v,t,t0,x,x0,m,P)

% Generate left boundary data for Legendre polynomials and uniform viscosity

% Indata:
% rho_0 — Solution scaling parameter (assumed deterministic)
% c1,c2 — Scaling parameters of uniform viscosity
% v — Advective velocity
% t — Time
% t0 — Initial time
% x — Vector of spatial grid points
% x0 — Initial pulse location
% m — Number of spatial grid points
% P — Number of gPC coefficients to be computed

% Outdata:
% g0 — Dirichlet data, left boundary



u_init = zeros(m*P,1);

%Generate normalized Legendre polynomials recursively

basis_fun = cell(1,P);
basis_fun{1} = @(xi) xi.^0;
basis_fun{2} = @(xi) sqrt(3)*xi;
for k=3:P
    basis_fun{k} = @(xi) (sqrt(2*k-3)/(k-1)*xi.*basis_fun{k-1}(xi)-(k-2)/((k-1)*sqrt(2*k-5))*basis_fun{k-2}(xi
        ))*sqrt(2*k-1);
end
```

```
mu_fun = @(xi) c1+c2*xi;

g0 = zeros(m*P,1);

for k=1:P
    integ = @(xi) 0.5.*basis_fun{k}(xi).*rho_0.*(1-erf((x(1)-(x0+v*(t+t0)))./sqrt(4*mu_fun(xi)*(t+t0))));
    g0(k,1) = quad(integ,-1,1);
end
```

C.1.3.2 bdy_cond_Legendre.m

```
function [g0 g1] = bdy_cond_Legendre(rho_0,c1,c2,v,t,t0,x,x0,m,P)

% Compute Dirichlet data for the left boundary and Neumann data for the
% right boundary, assuming Legendre polynomials representation

% Indata:
% rho_0 - Solution scaling parameter (assumed deterministic)
% c1,c2 - Scaling parameters of uniform viscosity
% v - Advective velocity
% t - Time
% t0 - Initial time
% x - Vector of spatial grid points
% x0 - Initial pulse location
% m - Number of spatial grid points
% P - Number of gPC coefficients to be computed

% Outdata:
% g0 - Dirichlet data, left boundary
% g1 - Neumann data, right boundary


u_init = zeros(m*P,1);

%Legendre polynomials

basis_fun = cell(1,P);
basis_fun{1} = @(xi) xi.^0;
basis_fun{2} = @(xi) sqrt(3)*xi;
for k=3:P
    basis_fun{k} = @(xi) (sqrt(2*k-3)/(k-1)*xi.*basis_fun{k-1}(xi)-(k-2)/((k-1)*sqrt(2*k-5))*basis_fun{k-2}(xi
        ))*sqrt(2*k-1);
end

mu_fun = @(xi) c1+c2*xi;

g0 = zeros(m*P,1);
g1 = zeros(m*P,1);

for k=1:P
    integ_0 = @(xi) 0.5*basis_fun{k}(xi).*rho_0./sqrt(4*pi*mu_fun(xi)*(t+t0)).*exp(-(x(1)-(x0+v*(t+t0))).^2.
        /(4*mu_fun(xi)*(t+t0)));
    g0(k,1) = quad(integ_0,-1,1);
    integ_1 = @(xi) 0.5*basis_fun{k}(xi).*rho_0./sqrt(4*pi*mu_fun(xi)*(t+t0)).*exp(-(x(end)-(x0+v*(t+t0))).^2.
        /(4*mu_fun(xi)*(t+t0))).*(-(x(end)-(x0+v*(t+t0)))./(2*mu_fun(xi)*(t+t0)));
    g1((m-1)*P+k,1) = quad(integ_1,-1,1);
end
```

C.1.3.3 bdy_cond_Hermite.m

```

function [g0 g1] = bdy_cond_Hermite(rho_0,c1,c2,v,t,t0,x,x0,m,P)

% Generate boundary data for Hermite polynomials and shifted lognormal viscosity

% Indata:
% rho_0 - Solution scaling parameter (assumed deterministic)
% c1,c2 - Scaling parameters of shifted lognormal viscosity
% v - Advective velocity
% t - Time
% t0 - Initial time
% x - Vector of spatial grid points
% x0 - Initial pulse location
% m - Number of spatial grid points
% P - Number of gPC coefficients to be computed

```

```
% Outdata:
% g0 - Dirichlet data, left boundary
% g1 - Neumann data, right boundary


tol = 15; % Replace infinite integration limit by sufficiently large real number

g0 = zeros(m*P,1);
g1 = zeros(m*P,1);

% Recursively generate Hermite polynomials
basis_fun = cell(1,P);
basis_fun{1} = @(xi) xi.^0;
basis_fun{2} = @(xi) xi;
for k=3:P
    basis_fun{k} = @(xi) 1/sqrt(k-1)*xi.*basis_fun{k-1}(xi)-sqrt((k-2)/(k-1))*basis_fun{k-2}(xi);
end

mu_fun = @(xi) c1+c2*exp(xi);

% Compute the gPC coefficients with numerical integration
for k=1:P
    integ = @(xi) 1/sqrt(2*pi).*exp(-xi.^2/2).*basis_fun{k}(xi).*rho_0./sqrt(4*pi*mu_fun(xi)*(t+t0)).*exp(-(x(1)-(x0+v*(t+t0))).^2./(4*mu_fun(xi).*(t+t0)));
    g0(k,1) = quad(integ,-tol,tol);

    integ = @(xi) -((x(end)-(x0+v*(t+t0)))./(2*mu_fun(xi)*(t+t0))).*(1/sqrt(2*pi).*exp(-xi.^2/2).*basis_fun{k}(xi).*rho_0./sqrt(4*pi*mu_fun(xi)*(t+t0)).*exp(-(x(end)-(x0+v*(t+t0))).^2./(4*mu_fun(xi).*(t+t0))));
    g1((m-1)*P+k,1) = quad(integ,-tol,tol);
end
```

C.1.4 Reference Solution

C.1.4.1 init_Hermite.m

```

function [u_init] = init_Hermite(rho_0,c1,c2,v,t,t0,x,x0,m,P)

% Generate initial function for Hermite polynomials and lognormal viscosity

% Indata:
% rho_0 - Solution scaling parameter (assumed deterministic)
% c1,c2 - Scaling parameters of lognormal viscosity
% v - Advective velocity
% t - Time
% t0 - Initial time
% x - Vector of spatial grid points
% x0 - Initial shock location
% m - Number of spatial grid points
% P - Number of gPC coefficients to be computed

% Outdata:
% u_init - gPC coefficients of the initial function evaluated at the spatial grid points


u_init = zeros(m*P,1);
tol = 20; % Replace integration limits of infinity by some sufficiently large number

% Generate normalized Hermite polynomials recursively

basis_fun = cell(1,P);
basis_fun{1} = @(xi) xi.^0;
basis_fun{2} = @(xi) xi;
for k=3:P
    basis_fun{k} = @(xi) 1/sqrt(k-1)*xi.*basis_fun{k-1}(xi)-sqrt((k-2)/(k-1))*basis_fun{k-2}(xi);
end

% Viscocity with shifted lognormal distribution
mu_fun = @(xi) c1+c2*exp(xi);

for k=1:P
    for j=1:m
        integ = @(xi) 1/sqrt(2*pi).*exp(-xi.^2/2).*basis_fun{k}(xi).*rho_0./sqrt(4*pi*mu_fun(xi)*(t+t0)).*exp(-(x(j)-(x0+v*(t+t0))).^2./(4*mu_fun(xi).*(t+t0)));
        u_init((j-1)*P+k,1) = quadgk(integ,-tol,tol); % Integration over the real line replaced with finite interval
    end
end
```

C.1.4.2 init_Legendre.m

```

function [u_init] = init_Legendre(rho_0,c1,c2,v,t,t0,x,x0,m,P)

% Generate initial function for Hermite polynomials and uniformly distributed viscosity

% Indata:
% rho_0 - Solution scaling parameter (assumed deterministic)
% c1,c2 - Scaling parameters of uniformly distributed viscosity
% v - Advective velocity
% t - Time
% t0 - Initial time
% x - Vector of spatial grid points
% x0 - Initial pulse location
% m - Number of spatial grid points
% P - Number of gPC coefficients to be computed

% Outdata:
% u_init - gPC coefficients of the initial function evaluated at the spatial grid points


u_init = zeros(m*P,1);

%Generate normalized Legendre polynomials recursively

basis_fun = cell(1,P);
basis_fun{1} = @(xi) xi.^0;
basis_fun{2} = @(xi) sqrt(3)*xi;
for k=3:P
    basis_fun{k} = @(xi) (sqrt(2*k-3)/(k-1)*xi.*basis_fun{k-1}(xi)-(k-2)/((k-1)*sqrt(2*k-5))*basis_fun{k-2}(xi))*sqrt(2*k-1);
end


mu_fun = @(xi) c1+c2*xi;

for k=1:P
    for j=1:m
        integ = @(xi) 0.5.*basis_fun{k}(xi).*rho_0./sqrt(4*pi*mu_fun(xi)*(t+t0)).*exp(-(x(j)-(x0+v*(t+t0))).^2./(4*mu_fun(xi).*(t+t0)));
        u_init((j-1)*P+k,1) = quad(integ,-1,1);
    end
end
```

C.1.4.3 statistics_adv_diff_lognorm.m

```

function [u_mean u_var] = statistics_adv_diff_lognorm(m,t,x,x0,c1,c2,rho_0,t0,v)

% Compute mean and variance for the solution assuming lognormal viscosity

% Indata:
% m - Number of spatial grid points
% t - Time
% x - Vector of spatial grid points
% x0 - Initial pulse location
% c1,c2 - Scaling parameters of lognormal viscosity
% rho_0 - Solution scaling parameter (assumed deterministic)
% t0 - Initial time
% v - Advective velocity

% Outdata:
% u_mean - Mean solution evaluated on the spatial grid
% u_var - Variance of the solution evaluated on the spatial grid


mu0 = @(xi) c1+c2*exp(xi); % Shifted lognormal model for the viscosity (xi standard Gaussian)
tol = 20; % Threshold for replacement of infinite integration limits

% For each spatial grid point, compute mean and variance
for j=1:m
    u = @(xi) rho_0./(4*pi*mu0(xi)*(t+t0)).^0.5.*exp(-(x(j)-(x0+v*(t+t0))).^2./(4*mu0(xi)*(t+t0)));
    u_mean(j,1) = quad(@(xi) exp(-xi.^2/2)/sqrt(2*pi).*u(xi),-tol,tol);
    u_var(j,1) = quad(@(xi) exp(-xi.^2/2)/sqrt(2*pi).*(u(xi).^2-u_mean(j,1).^2),-tol,tol);
end
```

C.1.4.4 statistics_adv_diff_uniform_mu.m

```
function [u_mean u_var] = statistics_adv_diff_uniform_mu(m,t,x,x0,c1,c2,rho_0,t0,v)

% Compute mean and variance for the solution assuming uniformly distributed viscosity

% Indata:
% m — Number of spatial grid points
% t — Time
% x — Vector of spatial grid points
% x0 — Initial pulse location
% c1,c2 — Scaling parameters of uniformly distributed viscosity
% rho_0 — Solution scaling parameter (assumed deterministic)
% t0 — Initial time
% v — Advective velocity

% Outdata:
% u_mean — Mean solution evaluated on the spatial grid
% u_var — Variance of the solution evaluated on the spatial grid


mu0 = @(xi) c1+c2*xi;

% For each spatial grid point, compute mean and variance
for j=1:m
    u = @(xi) rho_0./(4*pi*mu0(xi)*(t+t0)).^0.5.*exp(-(x(j)-(x0+v*(t+t0))).^2./(4*mu0(xi)*(t+t0)));
    u_mean(j,1) = quad(@(xi) 0.5*u(xi),-1,1);
    u_var(j,1) = quad(@(xi) 0.5*(u(xi).^2-u_mean(j,1).^2),-1,1);
end
```

C.2 Non-linear Transport

C.2.1 Main Code

C.2.1.1 burgers_main.m

```


% Stochastic Galerkin formulation of Burgers' equation, one spatial dimension
% Finite difference discretization in space (summation by parts + SAT) for SG
% Weak imposition of boundary conditions
% Runge-Kutta (4th order accurate) in time
% The assigned fourth and second order dissipation operators => one-sided diff. op.


clear all;

left = 0; right = 1; % Spatial end points
p = 4;% Number of polynomial chaos terms (order+1)
m = 100; % Discretization points in space
T = 0.7; % End time of simulation
% Solution parameters
mean_left = 0.9;
mean_right = -1.1;
sig_h_left = 0.3; % Assumed to be positive for correct analytical solution
% Assume same standard deviation everywhere for comparison with analytical
% solution
x0 = 0.5; % Initial location of shock, x0 in [left, right]


dx = (right-left)/(m-1);
dt = 0.05*dx;

sig_h_right = sig_h_left;
sig_h = sig_h_left;

x = linspace(left,right,m);

% Analytical solution of the truncated problem for p=1,2, else the exact
% coefficients of the infinte expansion problem
if p == 1
    u_ref = exact_solution_determ(mean_left,mean_right,T,m,left,right,x0,x);
elseif p == 2
    u_ref = exact_solution_2x2(mean_left,mean_right,sig_h,T,m,left,right,x0,x);
else
    u_ref = exact_solution_p_inf(mean_left,mean_right,sig_h,T,m,left,right,x0,x,p);
end
```

```
I_p = eye(p);
I_m = eye(m);
old = 1;
new = 2;
u = zeros(p*m,2);
order = 4; % Order of accuracy of SBP operators - SHOULD MATCH DISSIPATION OPERATORS

%Compute inner triple products
% Here we assume Gaussian distribution of the states, which causes no
% problem in terms of numerical stability
[C] = Hermite_chaos(p-1);

%Difference operators (D1 1st derivative)
[D1 D2,BS,S,P] = SBP_operators(m,dx,order);

% Invert the norm matrix P (assuming it is diagonal)
for j=1:m
    P_temp(j,j) = 1/P(j,j);
end
P_inv=sparse(P_temp);

kron_Dx = kron(D1,I_p);

%Initialization
u(:,old) = initial_conditions(m,p,C,x0,mean_left,sig_h_left,mean_right,sig_h_right,left,right);

%Compute force terms
E_1 = sparse(zeros(m));
E_1(1,1) = 1;
E_m = sparse(zeros(m));
E_m(m,m)=1;
[Sig_left,Sig_right] = penalty(p,u(1:p,1),u(p*(m-1)+1:p*m,1),C);
force_left = kron(I_m,I_p)*kron(P_inv,I_p)*kron(E_1,Sig_left);
force_right = kron(I_m,I_p)*kron(P_inv,I_p)*kron(E_m,Sig_right);

% Time dependent boundary conditions
t_tol = 1e-5; % Tolerance to avoid division with zero at t=0
if p == 1
    [g_left g_right] = boundary_cond_determ(mean_left,mean_right,t_tol,x0,left,right,m);
elseif p == 2
    [g_left g_right] = boundary_cond_2x2(mean_left,mean_right,sig_h,t_tol,x0,left,right,m);
else
    [g_left g_right] = boundary_cond_p_inf(mean_left,mean_right,sig_h,t_tol,x0,left,right,p,m);
end

%Iterate over time with fourth order Runge-Kutta
t=0;
while t<T
   if T-t<dt
       dt = T-t; % Make sure we end at t=T
    end
    t = t+dt;% Update the time

    % Compute system eigenvalues for dissipation. May be replaced by the
    % approximate expression in Chapter 6.

    max_ev = 0;
    for i=0:length(u)/p-1
        u_part = u(p*i+1:p*(i+1),1);
        eig_temp = max(abs(eig(I_p*A_matrix(u_part,p,C))));
        if eig_temp>max_ev
            max_ev = eig_temp;
        end
    end

    % These dissipation operators are chosen for the 4th order operators -
    % need to be adapted for different order of accuracy

    diss_2nd_ord = dissipation_2nd_der(m,p,P_inv,I_p,2*max_ev/(6*dx),dx);
    diss_4th_ord = dissipation_4th_der(m,p,P_inv,I_p,2*max_ev/(24*dx),dx);

    % Time-stepping by 4th order Runge-Kutta

    F1 = flux_func(u(:,old),p,C);
    k1 = dt*(-kron_Dx*F1+(diss_4th_ord+diss_2nd_ord+force_left+force_right)*u(:,old)-force_left*g_left-
        force_right*g_right);

    F2 = flux_func(u(:,old)+k1/2,p,C);
    k2 = dt*(-kron_Dx*F2+(diss_4th_ord+diss_2nd_ord+force_left+force_right)*(u(:,old)+k1/2)-force_left*g_left-
        force_right*g_right);

    F3 = flux_func(u(:,old)+k2/2,p,C);
    k3 = dt*(-kron_Dx*F3+(diss_4th_ord+diss_2nd_ord+force_left+force_right)*(u(:,old)+k2/2)-force_left*g_left-
        force_right*g_right);

    F4 = flux_func(u(:,old)+k3,p,C);
```

```
    k4 = dt*(-kron_Dx*F4+(diss_4th_ord+diss_2nd_ord+force_left+force_right)*(u(:,old)+k3)-force_left*g_left-
       force_right*g_right);

    u(:,new) = u(:,old)+1/6*(k1+2*k2+2*k3+k4);
    u(:,old) = u(:,new);

    % Update penalties for imposition of boundary conditions
    [Sig_left, Sig_right] = penalty(p,u(1:p,1),u(p*(m-1)+1:p*m,1),C);
    force_left = kron(I_m,I_p)*kron(P_inv,I_p)*kron(E_1,Sig_left);
    force_right = kron(I_m,I_p)*kron(P_inv,I_p)*kron(E_m,Sig_right);

    % Update the time dependent boundary conditions

    if p == 1
        [g_left g_right] = boundary_cond_determ(mean_left,mean_right,t,x0,left,right,m);
    elseif p == 2
        [g_left g_right] = boundary_cond_2x2(mean_left,mean_right,sig_h,t,x0,left,right,m);
    else
        [g_left g_right] = boundary_cond_p_inf(mean_left,mean_right,sig_h,t,x0,left,right,p,m);
    end

    % Plot the coefficients

    for k=1:p
        u_plot(:,k) = u(k:p:end,new);
    end
    plot(x,u_plot,'-*');
    title(['gPC coefficients, t=' num2str(t,'%.2f')])
    leg_strs = {};
    for k=1:p
        legstrs{k} = ['gPC coe. ' num2str(k-1)];
    end
    legend(legstrs);
    drawnow;
end

u=u(:,new);

for i=1:m
    u_var(i,1) = sum(u((i-1)*p+2:i*p,old).^2);
end


% Plot the numerical and analytical solution for the truncated 2x2 problem

if p == 2
    subplot(1,2,1);
    plot(x,u(1:p:end),'-','LineWidth',2,'Color','r');
    hold on;
    plot(x,u_ref(:,1),'--','LineWidth',2,'Color','b')
    legend('Numerical','Analytical, 2x2');
    title('u_0','fontsize',14,'fontweight','b');

    subplot(1,2,2);
    plot(x,u(2:p:(m-1)*p+2),'-','LineWidth',2,'Color','r');
    hold on;
    plot(x,u_ref(:,2),'--','LineWidth',2,'Color','b')
    legend('Numerical','Analytical, 2x2');
    title('u_1','fontsize',14,'fontweight','b');
end

% Plot the numerical approximation of order p and the analytical
% coefficients

if p ~= 2
    for k=1:p
        figure;
        plot(x,u(k:p:end),'-r',x,u_ref(:,k),'--b','LineWidth',2)
        legend('Numerical','Analytical');%,'Analytical expected value');
        title(['Solution gPC coefficient ' num2str(k-1)],'fontsize',14,'fontweight','b');
    end
end
```

C.2.1.2 Hermite_chaos.m

```

function [C] = Hermite_chaos(n)

% Compute hermite chaos parameters

% Indata:
% n - Order of gPC

% Outdata:
% C - Three term inner products C(i,j,k) = E[Phi_i Phi_j Phi_k]
```

```


C = zeros(n+1,n+1,n+1);
for i = 0:n
    for j = 0:n
        for k = 0:n
            s = (i+k+j) / 2;
            if rem(i+k+j,2) == 1 || i > s || j > s || k > s
                C(i+1,j+1,k+1) = 0;
            else
                C(i+1,j+1,k+1) = factorial(i) * factorial(j) ...
                    *factorial(k) / ((factorial(s-i) * factorial(s-j) * factorial(s-k)) * sqrt(factorial(i)) *
        sqrt(factorial(j)) * sqrt(factorial(k)));
            end
        end
    end
end
return
```

C.2.1.3 A_matrix.m

```

function [A] = A_matrix(u_loc,p,C)

% Compute the matrix A(u) of triple inner products,
% where A_{i,j} = sum_{k=0}^{p} int u_k psi_i psi_j psi_k dP

% Indata:
% u_loc - Vector of gPC coefficients of the argument u
% p - Number of gPC basis functions
% C - Precomputed inner triple products of the basis functions psi

% Outdata:
% A - matrix of sums of inner products


A = zeros(p);

for j=1:p
    A = A + C(:,:,j)*u_loc(j);
end
```

C.2.2 Discretization Operators

C.2.2.1 SBP_operators.m

```

function [D1 D2,BS,S,H] = SBP_operators(n,dx,order)

% SBP operators of orders 2, 4, 6 and 8 for the first and second derivative.

% Indata:
% n - Number of spatial grid pts
% dx - Step size
% order - Order of accuracy (only for 2,4,6,8)

% Outdata:
% D1 - First derivative operator
% D2 - Second derivative operator (D = P^{-1}M)
% S - First derivative operator on boundaries
% BS - The boundary elements in the energy estimate
% H - The norm operator (denoted P in some papers)


e = ones(n,1);

if order==2

    D1 = 1/dx*spdiags([-1/2*e 0*e 1/2*e],-1:1,n,n);
    D1(1,1) = -1/dx;
    D1(1,2) = 1/dx;
    D1(1,3) = 0;
    D1(n,n) = 1/dx;
    D1(n,n-1) = -1/dx;
    D1(n,n-2) = 0;

```

```
    %%%%%%
    D2 = 1/(dx^2)*spdiags([1*e -2*e 1*e],-1:1,n,n);
    D2(1,1) = 1/(dx^2);
    D2(1,2) = -2/(dx^2);
    D2(1,3) = 1/(dx^2);
    D2(n,n) = 1/(dx^2);
    D2(n,n-1) = -2/(dx^2);
    D2(n,n-2) = 1/(dx^2);


    H = dx*spdiags([e],0,n,n);
    H(1,1) = dx*1/2;
    H(n,n) = dx*1/2;

    BS = (1/dx)*spdiags(zeros(size(e)),0,n,n);
    BS(1,1) = 3/2/dx;
    BS(1,2) = -2/dx;
    BS(1,3) = 1/2/dx;
    BS(n,n) = 3/2/dx;
    BS(n,n-1) = -2/dx;
    BS(n,n-2) = 1/2/dx;

    S = (1/dx)*spdiags([e],0,n,n);
    S(1,1) = -3/2/dx;
    S(1,2) = 2/dx;
    S(1,3) = -1/2/dx;

    S(n,n) = 3/2/dx;
    S(n,n-1) = -2/dx;
    S(n,n-2) = 1/2/dx;

elseif order==4

    D1 = 1/dx*spdiags([1/12*e -2/3*e 0*e 2/3*e -1/12*e],-2:2,n,n);
    D1(1,1) = -24/17/dx;
    D1(1,2) = 59/34/dx;
    D1(1,3) = -4/17/dx;
    D1(1,4) = -3/34/dx;
    D1(1,5) = 0;
    D1(1,6) = 0;
    D1(2,1) = -1/2/dx;
    D1(2,2) = 0;
    D1(2,3) = 1/2/dx;
    D1(2,4:6) = 0;
    D1(3,1) = 4/43/dx;
    D1(3,2) = -59/86/dx;
    D1(3,3) = 0;
    D1(3,4) = 59/86/dx;
    D1(3,5) = -4/43/dx;
    D1(3,6) = 0;
    D1(4,1) = 3/98/dx;
    D1(4,2) = 0;
    D1(4,3) = -59/98/dx;
    D1(4,4) = 0;
    D1(4,5) = 32/49/dx;
    D1(4,6) = -4/49/dx;
    D1(4,7) = 0;
    D1(n,n) = -D1(1,1);
    D1(n,n-1) = -D1(1,2);
    D1(n,n-2) = -D1(1,3);
    D1(n,n-3) = -D1(1,4);
    D1(n,n-4) = -D1(1,5);
    D1(n,n-5) = -D1(1,6);
    D1(n-1,n) = -D1(2,1);
    D1(n-1,n-1) = -D1(2,2);
    D1(n-1,n-2) = -D1(2,3);
    D1(n-1,n-3) = -D1(2,4);
    D1(n-1,n-4) = -D1(2,5);
    D1(n-1,n-5) = -D1(2,6);
    D1(n-2,n) = -D1(3,1);
    D1(n-2,n-1) = -D1(3,2);
    D1(n-2,n-2) = -D1(3,3);
    D1(n-2,n-3) = -D1(3,4);
    D1(n-2,n-4) = -D1(3,5);
    D1(n-2,n-5) = -D1(3,6);
    D1(n-3,n) = -D1(4,1);
    D1(n-3,n-1) = -D1(4,2);
    D1(n-3,n-2) = -D1(4,3);
    D1(n-3,n-3) = -D1(4,4);
    D1(n-3,n-4) = -D1(4,5);
    D1(n-3,n-5) = -D1(4,6);


    %%%%%%
    D2 = 1/(dx^2)*spdiags([-1/12*e 4/3*e -5/2*e 4/3*e -1/12*e],-2:2,n,n);
    D2(1,1) = 2/(dx^2);
    D2(1,2) = -5/(dx^2);
    D2(1,3) = 4/(dx^2);
    D2(1,4) = -1/(dx^2);
```

```
    D2(2,1) = 1/(dx^2);
    D2(2,2) = -2/(dx^2);
    D2(2,3) = 1/(dx^2);
    D2(2,4) = 0;
    D2(3,1) = -4/43/(dx^2);
    D2(3,2) = 59/43/(dx^2);
    D2(3,3) = -110/43/(dx^2);
    D2(3,4) = 59/43/(dx^2);
    D2(3,5) = -4/43/(dx^2);
    D2(4,1) = -1/49/(dx^2);
    D2(4,2) = 0;
    D2(4,3) = 59/49/(dx^2);
    D2(4,4) = -118/49/(dx^2);
    D2(4,5) = 64/49/(dx^2);
    D2(4,6) = -4/49/(dx^2);
    D2(n,n) = D2(1,1);
    D2(n,n-1) = D2(1,2);
    D2(n,n-2) = D2(1,3);
    D2(n,n-3) = D2(1,4);
    D2(n-1,n) = D2(2,1);
    D2(n-1,n-1) = D2(2,2);
    D2(n-1,n-2) = D2(2,3);
    D2(n-1,n-3) = D2(2,4);
    D2(n-2,n) = D2(3,1);
    D2(n-2,n-1) = D2(3,2);
    D2(n-2,n-2) = D2(3,3);
    D2(n-2,n-3) = D2(3,4);
    D2(n-2,n-4) = D2(3,5);
    D2(n-3,n) = D2(4,1);
    D2(n-3,n-1) = D2(4,2);
    D2(n-3,n-2) = D2(4,3);
    D2(n-3,n-3) = D2(4,4);
    D2(n-3,n-4) = D2(4,5);
    D2(n-3,n-5) = D2(4,6);

    H = dx*spdiags(e,0,n,n);
    H(1,1) = dx*17/48;
    H(2,2) = dx*59/48;
    H(3,3) = dx*43/48;
    H(4,4) = dx*49/48;
    H(n,n) = H(1,1);
    H(n-1,n-1) = H(2,2);
    H(n-2,n-2) = H(3,3);
    H(n-3,n-3) = H(4,4);

    S = (1/dx)*spdiags(e,0,n,n);
    S(1,1) = -11/6/dx;
    S(1,2) = 3/dx;
    S(1,3) = -3/2/dx;
    S(1,4) = 1/3/dx;
    S(n,n) = 11/6/dx;
    S(n,n-1) = -3/dx;
    S(n,n-2) = 3/2/dx;
    S(n,n-3) = -1/3/dx;

    BS = (1/dx)*spdiags(zeros(size(e)),0,n,n);
    BS(1,1) = 11/6/dx;
    BS(1,2) = -3/dx;
    BS(1,3) = 3/2/dx;
    BS(1,4) = -1/3/dx;
    BS(n,n) = 11/6/dx;
    BS(n,n-1) = -3/dx;
    BS(n,n-2) = 3/2/dx;
    BS(n,n-3) = -1/3/dx;

elseif order==6
    e = ones(n,1);
    %%%%
    D1 = (1/(dx))*spdiags([-1/60*e 3/20*e -3/4*e 0*e 3/4*e -3/20*e 1/60*e],-3:3,n,n);

    D1(1,1) = -21600/13649/dx;
    D1(1,2) = 104009/54596/dx;
    D1(1,3) = 30443/81894/dx;
    D1(1,4) = -33311/27298/dx;
    D1(1,5) = 16863/27298/dx;
    D1(1,6) = -15025/163788/dx;
    D1(1,7) = 0;
    D1(1,8) = 0;
    D1(2,1) = -104009/240260/dx;
    D1(2,2) = 0;
    D1(2,3) = -311/72078/dx;
    D1(2,4) = 20229/24026/dx;
    D1(2,5) = -24337/48052/dx;
    D1(2,6) = 36661/360390/dx;
    D1(2,7) = 0;
    D1(2,8) = 0;
    D1(3,1) = -30443/162660/dx;
    D1(3,2) = 311/32532/dx;
    D1(3,3) = 0;
```

```
D1(3,4) = -11155/16266/dx;
D1(3,5) = 41287/32532/dx;
D1(3,6) = -21999/54220/dx;
D1(3,7) = 0;
D1(3,8) = 0;
D1(4,1) = 33311/107180/dx;
D1(4,2) = -20229/21436/dx;
D1(4,3) = 485/1398/dx;
D1(4,4) = 0;
D1(4,5) = 4147/21436/dx;
D1(4,6) = 25427/321540/dx;
D1(4,7) = 72/5359/dx;
D1(4,8) = 0;
D1(5,1) = -16863/78770/dx;
D1(5,2) = 24337/31508/dx;
D1(5,3) = -41287/47262/dx;
D1(5,4) = -4147/15754/dx;
D1(5,5) = 0;
D1(5,6) = 342523/472620/dx;
D1(5,7) = -1296/7877/dx;
D1(5,8) = 144/7877/dx;
D1(5,9) = 0;
D1(6,1) = 15025/525612/dx;
D1(6,2) = -36661/262806/dx;
D1(6,3) = 21999/87602/dx;
D1(6,4) = -25427/262806/dx;
D1(6,5) = -342523/525612/dx;
D1(6,6) = 0;
D1(6,7) = 32400/43801/dx;
D1(6,8) = -6480/43801/dx;
D1(6,9) = 720/43801/dx;
D1(6,10) = 0;
D1(n,n) = -D1(1,1);
D1(n,n-1) = -D1(1,2);
D1(n,n-2) = -D1(1,3);
D1(n,n-3) = -D1(1,4);
D1(n,n-4) = -D1(1,5);
D1(n,n-5) = -D1(1,6);
D1(n,n-6) = -D1(1,7);
D1(n,n-7) = -D1(1,8);
D1(n-1,n) = -D1(2,1);
D1(n-1,n-1) = -D1(2,2);
D1(n-1,n-2) = -D1(2,3);
D1(n-1,n-3) = -D1(2,4);
D1(n-1,n-4) = -D1(2,5);
D1(n-1,n-5) = -D1(2,6);
D1(n-1,n-6) = -D1(2,7);
D1(n-1,n-7) = -D1(2,8);
D1(n-2,n) = -D1(3,1);
D1(n-2,n-1) = -D1(3,2);
D1(n-2,n-2) = -D1(3,3);
D1(n-2,n-3) = -D1(3,4);
D1(n-2,n-4) = -D1(3,5);
D1(n-2,n-5) = -D1(3,6);
D1(n-2,n-6) = -D1(3,7);
D1(n-2,n-7) = -D1(3,8);
D1(n-3,n) = -D1(4,1);
D1(n-3,n-1) = -D1(4,2);
D1(n-3,n-2) = -D1(4,3);
D1(n-3,n-3) = -D1(4,4);
D1(n-3,n-4) = -D1(4,5);
D1(n-3,n-5) = -D1(4,6);
D1(n-3,n-6) = -D1(4,7);
D1(n-3,n-7) = -D1(4,8);
D1(n-4,n) = -D1(5,1);
D1(n-4,n-1) = -D1(5,2);
D1(n-4,n-2) = -D1(5,3);
D1(n-4,n-3) = -D1(5,4);
D1(n-4,n-4) = -D1(5,5);
D1(n-4,n-5) = -D1(5,6);
D1(n-4,n-6) = -D1(5,7);
D1(n-4,n-7) = -D1(5,8);
D1(n-4,n-8) = -D1(5,9);
D1(n-5,n) = -D1(6,1);
D1(n-5,n-1) = -D1(6,2);
D1(n-5,n-2) = -D1(6,3);
D1(n-5,n-3) = -D1(6,4);
D1(n-5,n-4) = -D1(6,5);
D1(n-5,n-5) = -D1(6,6);
D1(n-5,n-6) = -D1(6,7);
D1(n-5,n-7) = -D1(6,8);
D1(n-5,n-8) = -D1(6,9);
D1(n-5,n-9) = -D1(6,10);

%%%%
D2 = (1/(dx^2))*spdiags([1/90*e -3/20*e 3/2*e -49/18*e 3/2*e -3/20*e 1/90*e],-3:3,n,n);

D2(1,1) = 114170/40947/(dx^2);
D2(1,2) = -438107/54596/(dx^2);
```

```
D2(1,3) = 336409/40947/(dx^2);
D2(1,4) = -276997/81894/(dx^2);
D2(1,5) = 3747/13649/(dx^2);
D2(1,6) = 21035/163788/(dx^2);
D2(1,7) = 0;
D2(1,8) = 0;
D2(2,1) = 6173/5860/(dx^2);
D2(2,2) = -2066/879/(dx^2);
D2(2,3) = 3283/1758/(dx^2);
D2(2,4) =-303/293/(dx^2);
D2(2,5) = 2111/3516/(dx^2);
D2(2,6) = -601/4395/(dx^2);
D2(2,7) = 0;
D2(2,8) = 0;
D2(3,1) = -52391/81330/(dx^2);
D2(3,2) = 134603/32532/(dx^2);
D2(3,3) = -21982/2711/(dx^2);
D2(3,4) = 112915/16266/(dx^2);
D2(3,5) = -46969/16266/(dx^2);
D2(3,6) = 30409/54220/(dx^2);
D2(3,7) = 0;
D2(3,8) = 0;
D2(4,1) = 68603/321540/(dx^2);
D2(4,2) = -12423/10718/(dx^2);
D2(4,3) = 112915/32154/(dx^2);
D2(4,4) = -75934/16077/(dx^2);
D2(4,5) = 53369/21436/(dx^2);
D2(4,6) = -54899/160770/(dx^2);
D2(4,7) = 48/5359/(dx^2);
D2(4,8) = 0;
D2(5,1) = -7053/39385/(dx^2);
D2(5,2) = 86551/94524/(dx^2);
D2(5,3) = -46969/23631/(dx^2);
D2(5,4) = 53369/15754/(dx^2);
D2(5,5) = -87904/23631/(dx^2);
D2(5,6) = 820271/472620/(dx^2);
D2(5,7) = -1296/7877/(dx^2);
D2(5,8) = 96/7877/(dx^2);
D2(5,9) = 0;
D2(6,1) = 21035/525612/(dx^2);
D2(6,2) = -24641/131403/(dx^2);
D2(6,3) = 30409/87602/(dx^2);
D2(6,4) = -54899/131403/(dx^2);
D2(6,5) = 820271/525612/(dx^2);
D2(6,6) = -117600/43801/(dx^2);
D2(6,7) = 64800/43801/(dx^2);
D2(6,8) = -6480/43801/(dx^2);
D2(6,9) = 480/43801/(dx^2);
D2(6,10) = 0;
D2(n,n) = D2(1,1);
D2(n,n-1) = D2(1,2);
D2(n,n-2) = D2(1,3);
D2(n,n-3) = D2(1,4);
D2(n,n-4) = D2(1,5);
D2(n,n-5) = D2(1,6);
D2(n,n-6) = D2(1,7);
D2(n,n-7) = D2(1,8);
D2(n-1,n) = D2(2,1);
D2(n-1,n-1) = D2(2,2);
D2(n-1,n-2) = D2(2,3);
D2(n-1,n-3) = D2(2,4);
D2(n-1,n-4) = D2(2,5);
D2(n-1,n-5) = D2(2,6);
D2(n-1,n-6) = D2(2,7);
D2(n-1,n-7) = D2(2,8);
D2(n-2,n) = D2(3,1);
D2(n-2,n-1) = D2(3,2);
D2(n-2,n-2) = D2(3,3);
D2(n-2,n-3) = D2(3,4);
D2(n-2,n-4) = D2(3,5);
D2(n-2,n-5) = D2(3,6);
D2(n-2,n-6) = D2(3,7);
D2(n-2,n-7) = D2(3,8);
D2(n-3,n) = D2(4,1);
D2(n-3,n-1) = D2(4,2);
D2(n-3,n-2) = D2(4,3);
D2(n-3,n-3) = D2(4,4);
D2(n-3,n-4) = D2(4,5);
D2(n-3,n-5) = D2(4,6);
D2(n-3,n-6) = D2(4,7);
D2(n-3,n-7) = D2(4,8);
D2(n-4,n) = D2(5,1);
D2(n-4,n-1) = D2(5,2);
D2(n-4,n-2) = D2(5,3);
D2(n-4,n-3) = D2(5,4);
D2(n-4,n-4) = D2(5,5);
D2(n-4,n-5) = D2(5,6);
D2(n-4,n-6) = D2(5,7);
D2(n-4,n-7) = D2(5,8);
```

```
    D2(n-4,n-8) = D2(5,9);
    D2(n-5,n) = D2(6,1);
    D2(n-5,n-1) = D2(6,2);
    D2(n-5,n-2) = D2(6,3);
    D2(n-5,n-3) = D2(6,4);
    D2(n-5,n-4) = D2(6,5);
    D2(n-5,n-5) = D2(6,6);
    D2(n-5,n-6) = D2(6,7);
    D2(n-5,n-7) = D2(6,8);
    D2(n-5,n-8) = D2(6,9);
    D2(n-5,n-9) = D2(6,10);

    H = dx*spdiags([e],0,n,n);

    H(1,1) = dx*13649/43200;
    H(2,2) = dx*12013/8640;
    H(3,3) = dx*2711/4320;
    H(4,4) = dx*5359/4320;
    H(5,5) = dx*7877/8640;
    H(6,6) = dx*43801/43200;
    H(n,n) = H(1,1);
    H(n-1,n-1) = H(2,2);
    H(n-2,n-2) = H(3,3);
    H(n-3,n-3) = H(4,4);
    H(n-4,n-4) = H(5,5);
    H(n-5,n-5) = H(6,6);

    BS = (1/dx)*spdiags([zeros(size(e))],0,n,n);

    BS(1,1) = 25/12/dx;
    BS(1,2) = -4/dx;
    BS(1,3) = 3/dx;
    BS(1,4) = -4/3/dx;
    BS(1,5) = 1/4/dx;
    BS(n,n) = BS(1,1);
    BS(n,n-1) = BS(1,2);
    BS(n,n-2) = BS(1,3);
    BS(n,n-3) = BS(1,4);
    BS(n,n-4) = BS(1,5);

    S = (1/dx)*spdiags([e],0,n,n);

    S(1,1) = -25/12/dx;
    S(1,2) = 4/dx;
    S(1,3) = -3/dx;
    S(1,4) = 4/3/dx;
    S(1,5) = -1/4/dx;
    S(n,n) = BS(1,1);
    S(n,n-1) = BS(1,2);
    S(n,n-2) = BS(1,3);
    S(n,n-3) = BS(1,4);
    S(n,n-4) = BS(1,5);
elseif order==8
    e = ones(n,1);

    D = (1/(dx^2))*spdiags([-1/560*e 8/315*e -1/5*e 8/5*e -205/72*e 8/5*e -1/5*e 8/315*e -1/560*e],-4:4,n,n);
    % eighth order standard central stencil

    D(1,1) = 4870382994799/1358976868290/(dx^2);
    D(1,2) = -893640087518/75498714905/(dx^2);
    D(1,3) = 926594825119/60398971924/(dx^2);
    D(1,4) = -1315109406200/135897686829/(dx^2);
    D(1,5) = 39126983272/15099742981/(dx^2);
    D(1,6) = 12344491342/75498714905/(dx^2);
    D(1,7) = -451560522577/2717953736580/(dx^2);
    D(1,8) = 0;
    D(1,9) = 0;
    D(1,10) = 0;
    D(1,11) = 0;
    D(1,12) = 0;
    D(2,1) = 333806012194/390619153855/(dx^2);
    D(2,2) = -154646272029/111605472530/(dx^2);
    D(2,3) = 1168338040/33481641759/(dx^2);
    D(2,4) = 82699112501/133926567036/(dx^2);
    D(2,5) = -171562838/11160547253/(dx^2);
    D(2,6) = -28244698346/167408208795/(dx^2);
    D(2,7) = 11904122576/167408208795/(dx^2);
    D(2,8) = -2598164715/312495323084/(dx^2);
    D(2,9) = 0;
    D(2,10) = 0;
    D(2,11) = 0;
    D(2,12) = 0;
    D(3,1) = 7838984095/52731029988/(dx^2);
    D(3,2) = 1168338040/5649753213/(dx^2);
    D(3,3) = -88747895/144865467/(dx^2);
    D(3,4) = 423587231/627750357/(dx^2);
    D(3,5) = -43205598281/22599012852/(dx^2);
    D(3,6) = 4876378562/1883251071/(dx^2);
    D(3,7) = -5124426509/3766502142/(dx^2);
```

```
D(3,8) = 10496900965/39548272491/(dx^2);
D(3,9) = 0;
D(3,10) = 0;
D(3,11) = 0;
D(3,12) = 0;
D(4,1) = -94978241528/828644350023/(dx^2);
D(4,2) = 82699112501/157837019052/(dx^2);
D(4,3) = 1270761693/13153084921/(dx^2);
D(4,4) = -167389605005/118377764289/(dx^2);
D(4,5) = 48242560214/39459254763/(dx^2);
D(4,6) = -31673996013/52612339684/(dx^2);
D(4,7) = 43556319241/118377764289/(dx^2);
D(4,8) = -44430275135/552429566682/(dx^2);
D(4,9) = 0;
D(4,10) = 0;
D(4,11) = 0;
D(4,12) = 0;
D(5,1) = 1455067816/21132528431/(dx^2);
D(5,2) = -171562838/3018932633/(dx^2);
D(5,3) = -43205598281/36227191596/(dx^2);
D(5,4) = 48242560214/9056797899/(dx^2);
D(5,5) = -52276055645/6037865266/(dx^2);
D(5,6) = 57521587238/9056797899/(dx^2);
D(5,7) = -80321706377/36227191596/(dx^2);
D(5,8) = 8078087158/21132528431/(dx^2);
D(5,9) = -1296/299527/(dx^2);
D(5,10) = 0;
D(5,11) = 0;
D(5,12) = 0;
D(6,1) = 10881504334/327321118845/(dx^2);
D(6,2) = -28244698346/140280479505/(dx^2);
D(6,3) = 4876378562/9352031967/(dx^2);
D(6,4) = -10557998671/12469375956/(dx^2);
D(6,5) = 57521587238/28056095901/(dx^2);
D(6,6) = -278531401019/93520319670/(dx^2);
D(6,7) = 73790130002/46760159835/(dx^2);
D(6,8) = -137529995233/785570685228/(dx^2);
D(6,9) = 2048/103097/(dx^2);
D(6,10) = -144/103097/(dx^2);
D(6,11) = 0;
D(6,12) = 0;
D(7,1) = -135555328849/8509847458140/(dx^2);
D(7,2) = 11904122576/101307707835/(dx^2);
D(7,3) = -5124426509/13507694378/(dx^2);
D(7,4) = 43556319241/60784624701/(dx^2);
D(7,5) = -80321706377/81046166268/(dx^2);
D(7,6) = 73790130002/33769235945/(dx^2);
D(7,7) = -950494905688/303923123505/(dx^2);
D(7,8) = 239073018673/141830790969/(dx^2);
D(7,9) = -145152/670091/(dx^2);
D(7,10) = 18432/670091/(dx^2);
D(7,11) = -1296/670091/(dx^2);
D(7,12) = 0;
D(8,1) = 0;
D(8,2) = -2598164715/206729925524/(dx^2);
D(8,3) = 10496900965/155047444143/(dx^2);
D(8,4) = -44430275135/310094888286/(dx^2);
D(8,5) = 425162482/2720130599/(dx^2);
D(8,6) = -137529995233/620189776572/(dx^2);
D(8,7) = 239073018673/155047444143/(dx^2);
D(8,8) = -144648000000/51682481381/(dx^2);
D(8,9) = 8128512/5127739/( dx^2);
D(8,10) = -1016064/5127739/(dx^2);
D(8,11) = 129024/5127739/( dx^2);
D(8,12) = -9072/5127739/(dx^2);

D(n,n) = D(1,1);
D(n,n-1) = D(1,2);
D(n,n-2) = D(1,3);
D(n,n-3) = D(1,4);
D(n,n-4) = D(1,5);
D(n,n-5) = D(1,6);
D(n,n-6) = D(1,7);
D(n,n-7) = D(1,8);
D(n,n-8) = D(1,9);
D(n,n-9) = D(1,10);
D(n,n-10) = D(1,11);
D(n,n-11) = D(1,12);
D(n-1,n) = D(2,1);
D(n-1,n-1) = D(2,2);
D(n-1,n-2) = D(2,3);
D(n-1,n-3) = D(2,4);
D(n-1,n-4) = D(2,5);
D(n-1,n-5) = D(2,6);
D(n-1,n-6) = D(2,7);
D(n-1,n-7) = D(2,8);
D(n-1,n-8) = D(2,9);
D(n-1,n-9) = D(2,10);
D(n-1,n-10) = D(2,11);
```

```
D(n-1,n-11) = D(2,12);
D(n-2,n) = D(3,1);
D(n-2,n-1) = D(3,2);
D(n-2,n-2) = D(3,3);
D(n-2,n-3) = D(3,4);
D(n-2,n-4) = D(3,5);
D(n-2,n-5) = D(3,6);
D(n-2,n-6) = D(3,7);
D(n-2,n-7) = D(3,8);
D(n-2,n-8) = D(3,9);
D(n-2,n-9) = D(3,10);
D(n-2,n-10) = D(3,11);
D(n-2,n-11) = D(3,12);
D(n-3,n) = D(4,1);
D(n-3,n-1) = D(4,2);
D(n-3,n-2) = D(4,3);
D(n-3,n-3) = D(4,4);
D(n-3,n-4) = D(4,5);
D(n-3,n-5) = D(4,6);
D(n-3,n-6) = D(4,7);
D(n-3,n-7) = D(4,8);
D(n-3,n-8) = D(4,9);
D(n-3,n-9) = D(4,10);
D(n-3,n-10) = D(4,11);
D(n-3,n-11) = D(4,12);
D(n-4,n) = D(5,1);
D(n-4,n-1) = D(5,2);
D(n-4,n-2) = D(5,3);
D(n-4,n-3) = D(5,4);
D(n-4,n-4) = D(5,5);
D(n-4,n-5) = D(5,6);
D(n-4,n-6) = D(5,7);
D(n-4,n-7) = D(5,8);
D(n-4,n-8) = D(5,9);
D(n-4,n-9) = D(5,10);
D(n-4,n-10) = D(5,11);
D(n-4,n-11) = D(5,12);
D(n-5,n) = D(6,1);
D(n-5,n-1) = D(6,2);
D(n-5,n-2) = D(6,3);
D(n-5,n-3) = D(6,4);
D(n-5,n-4) = D(6,5);
D(n-5,n-5) = D(6,6);
D(n-5,n-6) = D(6,7);
D(n-5,n-7) = D(6,8);
D(n-5,n-8) = D(6,9);
D(n-5,n-9) = D(6,10);
D(n-5,n-10) = D(6,11);
D(n-5,n-11) = D(6,12);
D(n-6,n) = D(7,1);
D(n-6,n-1) = D(7,2);
D(n-6,n-2) = D(7,3);
D(n-6,n-3) = D(7,4);
D(n-6,n-4) = D(7,5);
D(n-6,n-5) = D(7,6);
D(n-6,n-6) = D(7,7);
D(n-6,n-7) = D(7,8);
D(n-6,n-8) = D(7,9);
D(n-6,n-9) = D(7,10);
D(n-6,n-10) = D(7,11);
D(n-6,n-11) = D(7,12);
D(n-7,n) = D(8,1);
D(n-7,n-1) = D(8,2);
D(n-7,n-2) = D(8,3);
D(n-7,n-3) = D(8,4);
D(n-7,n-4) = D(8,5);
D(n-7,n-5) = D(8,6);
D(n-7,n-6) = D(8,7);
D(n-7,n-7) = D(8,8);
D(n-7,n-8) = D(8,9);
D(n-7,n-9) = D(8,10);
D(n-7,n-10) = D(8,11);
D(n-7,n-11) = D(8,12);


H = dx*spdiags([e],0,n,n);

H(1,1) = dx*1498139/5080320;
H(2,2) = dx*1107307/725760;
H(3,3) = dx*20761/80640;
H(4,4) = dx*1304999/725760;
H(5,5) = dx*299527/725760;
H(6,6) = dx*103097/80640;
H(7,7) = dx*670091/725760;
H(8,8) = dx*5127739/5080320;
H(n,n) = H(1,1);
H(n-1,n-1) = H(2,2);
H(n-2,n-2) = H(3,3);
H(n-3,n-3) = H(4,4);
```

```
    H(n-4,n-4) = H(5,5);
    H(n-5,n-5) = H(6,6);
    H(n-6,n-6) = H(7,7);
    H(n-7,n-7) = H(8,8);

    BS = (1/dx)*spdiags([zeros(size(e))],0,n,n);

    BS(1,1) = 4723/2100/dx;
    BS(1,2) = -839/175/dx;
    BS(1,3) = 157/35/dx;
    BS(1,4) = -278/105/dx;
    BS(1,5) = 103/140/dx;
    BS(1,6) = 1/175/dx;
    BS(1,7) = -6/175/dx;
    BS(n,n) = BS(1,1);
    BS(n,n-1) = BS(1,2);
    BS(n,n-2) = BS(1,3);
    BS(n,n-3) = BS(1,4);
    BS(n,n-4) = BS(1,5);
    BS(n,n-5) = BS(1,6);
    BS(n,n-6) = BS(1,7);


    S = (1/dx)*spdiags([e],0,n,n);

    S(1,1) = -4723/2100/dx;
    S(1,2) = 839/175/dx;
    S(1,3) = -157/35/dx;
    S(1,4) = 278/105/dx;
    S(1,5) = -103/140/dx;
    S(1,6) = -1/175/dx;
    S(1,7) = 6/175/dx;
    S(n,n) = BS(1,1);
    S(n,n-1) = BS(1,2);
    S(n,n-2) = BS(1,3);
    S(n,n-3) = BS(1,4);
    S(n,n-4) = BS(1,5);
    S(n,n-5) = BS(1,6);
    S(n,n-6) = BS(1,7);

else
    disp('Only order 2, 4, 6 or 8 implemented here.')
end
```

C.2.2.2 flux_func.m

```

function [flux] = flux_func(u,p,C)

% Flux function, yields the stochastic Galerkin flux f = 0.5*A(u)u

% Indata:
% u - Vector of solution variables (gPC coefficients)
% p - Number of gPC basis functions
% C - Triple product matrix

% Outdata:
% flux - Stochastic Galerkin flux function


flux = zeros(length(u),1);

for i=0:length(u)/p-1 % Loop over the spatial grid points
    u_part=u(p*i+1:p*(i+1),1);
    flux(i*p+1:(i+1)*p,1) = 0.5*A_matrix(u_part,p,C)*u_part;
end
```

C.2.2.3 dissipation_2nd_der.m

```

function [diss_op] = dissipation_2nd_der(m,p,P_inv,H_inv,const,dx)

% Dissipation operator corresponding to second derivative to a system of size m*p (space * PCE-coeff.)
% Global dissipation constant

% Indata:
% m - Number of spatial grid points
% p - Number of gPC coefficients
```

```
% P_inv — Inverse of SBP norm matrix P
% H_inv — Inverse of SG mass matrix (in our implementation it is always the identity matrix)
% const — Dissipation constant
% dx — Spatial grid size

% Outdata:
% diss_op — Discrete dissipation matrix

D = zeros(m)+diag(ones(m,1))—diag(ones(m—1,1),—1);
D(1,1) = —1;
D(1,2) = 1;
D = sparse(D);

B = const*eye(m*p);
B(1,1) = 0;
diss_op=—dx*kron(P_inv*D',eye(p))*B*kron(D,H_inv);
```

C.2.2.4 dissipation_4th_der.m

```

function [diss_op] = dissipation_4th_der(m,p,P2_inv,H_inv,const,dx)

% Dissipation operator corresponding to fourth order derivative to a system of size m*p (space * PCE—coeff.)
% Global dissipation constant

% Indata:
% m — Number of spatial grid points
% p — Number of gPC coefficients
% P2_inv — Inverse of SBP norm matrix P2
% H_inv — Inverse of SG mass matrix (in our implementation it is always the identity matrix)
% const — Dissipation constant
% dx — Spatial grid size

% Outdata:
% diss_op — Discrete dissipation matrix

dia=[1 —2 1];
D2=spdiags(ones(m,1)*dia,[—1:1],m,m);
D2(1,1:3)=dia;
D2(m,m—2:m)=dia;

B2 = const*eye(m);
B2(1,1) = 0;

diss_op=kron(—dx*P2_inv*D2'*B2*D2,H_inv);
```

C.2.3 Boundary Treatment

C.2.3.1 penalty.m

```

function [Sig_left,Sig_right]=penalty(p,u_bc_l,u_bc_r,C)

% Assign penalty matrix for conservative system

% Indata:
% p — Number of gPC coefficients (p=M—1)
% u_bc_l, u_bc_r — Left and right boundary values
% C — Matrices of inner triple products of gPC basis functions

% Outdata:
% Sig_left, Sig_right — Left and right penalty matrices (SAT)

if p>1
    A_l = (C(:,:,1)*u_bc_l(1)+C(:,:,2)*u_bc_l(2));
    A_r = (C(:,:,1)*u_bc_r(1)+C(:,:,2)*u_bc_r(2));
    %Decomposition of the system matrix according to the signs of the
    %eigenvalues
    [X_l,D_l] = eig(A_l);
    [X_r,D_r] = eig(A_r);
    for i=1:p
        if D_l(i,i)<0
            D_l(i,i) = 0;
        end
        if D_r(i,i)>0
```

```
            D_r(i,i) = 0;
        end
    end
    A_l = X_l*D_l*X_l';
    A_r = X_r*D_r*X_r';

    %Scaling with 0.5 for conservative systems
    Sig_left = -1/2*A_l;
    Sig_right = 1/2*A_r;

end
if p==1
    Sig_left = -1/2*u_bc_l(1);
    Sig_right = 1/2*u_bc_r(1);
end
```

C.2.3.2 boundary_cond_determ.m

```

function [g_left g_right] = boundary_cond_determ(mean_left,mean_right,t,x0,left,right,m)

% Compute boundary conditions for the deterministic Burgers' equation

% Indata:
% mean_left,mean_right - Left and right states
% t - Time
% x0 - Initial shock location
% left - Lower limit of spatial interval
% right - Upper limit of spatial interval
% m - Number of spatial grid points

% Outdata:
% g_left - Left boundary Dirichlet data
% g_right - Right boundary Dirichlet data


g_left = zeros(m,1);
g_right = zeros(m,1);

% Shock speed
s = (mean_left + mean_right)/2;

if x0+s*t < left
   g_left(1) = mean_right;
   g_right(end) = mean_right;
end

if x0+s*t >= left && x0+s*t <= right
   g_left(1) = mean_left;
   g_right(end) = mean_right;
end

if x0+s*t > right
   g_left(1) = mean_left;
   g_right(end) = mean_left;
end
```

C.2.3.3 boundary_cond_2x2.m

```

function [g_left g_right] = boundary_cond_2x2(mean_left,mean_right,sig_h,t,x0,left,right,m)

% Compute the boundary conditions of the 2x2 stochastic Galerkin form of
% Burgers' equation for the Riemann problem

% Indata:
% mean_left,mean_right - Mean (u_0) of the left and right states
% sig_h - Standard deviation (u_1), assumed uniform over the spatial domain
% t - Time
% x0 - Initial shock location
% left - Lower limit of spatial interval
% right - Upper limit of spatial interval
% m - Number of spatial grid points

% Outdata:
% g_left - Dirichlet condition for left boundary
% g_right - Dirichlet condition for right boundary

```

```

% Exact solution for the 2 x 2 case

s1 = (mean_left+mean_right)/2-sig_h; % Shock speed 1
s2 = (mean_left+mean_right)/2+sig_h; % Shock speed 2

g_left = zeros(2*m,1);
g_right = zeros(2*m,1);
g_left(1) = mean_left;
g_left(2) = sig_h;
g_right(2*(m-1)+1) = mean_right;
g_right(2*(m-1)+2) = sig_h;

% One wave propagating to the left, the other to the right
if s1<=0 && s2>=0
    if t>(left-x0)/s1
        g_left(1)=(mean_left+mean_right)/2
        g_left(2) = (mean_left-mean_right)/2+sig_h;

    end
    if t>(right-x0)/s2
        g_right(2*(m-1)+1) = (mean_left+mean_right)/2;
        g_right(2*(m-1)+2) = (mean_left-mean_right)/2+sig_h;
    end
end

% Both waves propagating to the left
if s1<=0 && s2<=0
    if t>(left-x0)/s1 && t<(left-x0)/s2
        g_left(1)=(mean_left+mean_right)/2;
        g_left(2) = (mean_left-mean_right)/2+sig_h;

    end
    if t>(left-x0)/s2
        g_left(1) = mean_right;
        g_left(2) = sig_h;
    end
end

% Both waves propagating to the right
if s1>=0 && s2>=0
    if t>(right-x0)/s2 && t < (right-x0)/s1
        g_right(2*(m-1)+1) = (mean_left+mean_right)/2;
        g_right(2*(m-1)+2) = (mean_left-mean_right)/2+sig_h;
    end
    if t > (right-x0)/s1
        g_right(2*(m-1)+1) = mean_left;
        g_right(2*(m-1)+2) = sig_h;
    end
end
```

C.2.3.4 boundary_cond_p_inf.m

```

function [g_left g_right] = boundary_cond_p_inf(mean_left,mean_right,sig_h,t,x0,left,right,p,m)

% Calculate time dependent boundary conditions for the first p coefficients of the infinite order expansion

% Indata:
% mean_left,mean_right - Left and right states
% sig_h - Standard deviation (uniform in space)
% t - Time
% x0 - Initial shock location
% left - Lower limit of spatial interval
% right - Upper limit of spatial interval
% p - Number of gPC coefficients to be computed
% m - Number of spatial grid points

% Outdata:
% g_left - Left boundary Dirichlet data for the vector of gPC coefficients
% g_right - Right boundary Dirichlet data for the vector of gPC coefficients


g_left = zeros(p*m,1);
g_right = zeros(p*m,1);

xi_l = (left-x0)./(sig_h*t)-(mean_left+mean_right)/(2*sig_h);
xi_r = (right-x0)./(sig_h*t)-(mean_left+mean_right)/(2*sig_h);
g_left(1) = mean_left + (mean_right-mean_left)*normcdf(xi_l,0,1);
g_right((m-1)*p+1) = mean_left + (mean_right-mean_left)*normcdf(xi_r,0,1);

```

```
g_left(2) = sig_h + (mean_left-mean_right)*exp(-xi_l^2/2)/sqrt(2*pi);
g_right((m-1)*p+2) = sig_h + (mean_left-mean_right)*exp(-xi_r^2/2)/sqrt(2*pi);
Psi_l(1:2) = [1 xi_l];
Psi_r(1:2) = [1 xi_r];

for k=3:p
    Psi_l(k) = xi_l.*sqrt(factorial(k-2)/factorial(k-1))*Psi_l(k-1) - (k-2)*sqrt(factorial(k-3)/factorial(k-1)
       ).*Psi_l(k-2);
    Psi_r(k) = xi_r.*sqrt(factorial(k-2)/factorial(k-1))*Psi_r(k-1) - (k-2)*sqrt(factorial(k-3)/factorial(k-1)
       ).*Psi_r(k-2);

    g_left(k) = (mean_left-mean_right)/sqrt(k-1)*exp(-xi_l^2/2)/sqrt(2*pi).*Psi_l(k-1);
    g_right((m-1)*p+k) = (mean_left-mean_right)/sqrt(k-1)*exp(-xi_r^2/2)/sqrt(2*pi).*Psi_r(k-1);
end
```

C.2.4 Reference Solution

C.2.4.1 initial_conditions.m

```

function [u_init] = initial_conditions(m,p,C,x0,mean_left,std_left,mean_right,std_right,left,right)

% Compute initial conditions (gPC coefficients) for the Riemann problem

% Indata:
% m - Number of spatial grid points
% p - Number of gPC coefficients to be computed
% C - Inner triple product matrices
% x0 - Initial shock location
% mean_left - Left mean state
% std_left - Standard deviation left state
% mean_right - Right mean state
% std_right - Standard deviation right state
% left - Lower limit of spatial interval
% right - Upper limit of spatial interval

% Output:
% u_init - Vector of initial gPC coefficients

u_init = zeros(m*p,1);

for i=1:p:p*(ceil(m*(x0-left)/(right-left))-1)+1
    u_init(i) = mean_left;
    if p>1
        u_init(i+1) = std_left;
    end
end
for (i=ceil(m*(x0-left)/(right-left))*p+1:p:p*(m-1)+1)
    u_init(i) = mean_right;
    if p>1
        u_init(i+1) = std_right;
    end
end
```

C.2.4.2 exact_solution_2x2.m

```

function [u_ref] = exact_solution_2x2(mean_left,mean_right,sig_h,T,m,left,right,x0,x)

% Compute the analytical solution of the 2x2 stochastic Galerkin Burgers'
% equation

% Indata:
% mean_left,mean_right - Mean (u_0) of the left and right states
% sig_h - Standard deviation (u_1), assumed uniform over the spatial domain
% T - Time
% m - Number of spatial grid points
% left - Lower limit of spatial interval
% right - Upper limit of spatial interval
% x0 - Initial shock location
% x - Vector of spatial grid points

% Outdata:
% u_ref - Analysical solution

```

```
s1 = (mean_left+mean_right)/2-sig_h; % Shock speed 1
s2 = (mean_left+mean_right)/2+sig_h; % Shock speed 2

u_ref = zeros(m,2);

for j=1:m
    if x(j)< x0+s1*T
        u_ref(j,1)=mean_left;
        u_ref(j,2)=sig_h;
    end
    if x(j)>= x0+s1*T && x(j) < x0+s2*T
        u_ref(j,1)= (mean_left+mean_right)/2;
        u_ref(j,2) = (mean_left-mean_right)/2+sig_h;
    end
    if x(j)> x0+s2*T
        u_ref(j,1)= mean_right;
        u_ref(j,2) = sig_h;
    end
end
```

C.2.4.3 exact_solution_determ.m

```
function [u_ref] = exact_solution_determ(mean_left,mean_right,T,m,left,right,x0,x)

% Compute the exact solution of the deterministic Burgers' equation

% Indata:
% mean_left,mean_right - Left and right states
% T - Time
% m - Number of spatial grid points
% left - Lower limit of spatial interval
% right - Upper limit of spatial interval
% x0 - Initial shock location
% x - Vector of spatial grid points

% Outdata:
% u_ref - Analytical solution


s = (mean_left + mean_right)/2; % Shock speed
u_ref = zeros(m,1);

for j=1:m
    if x(j)< x0+s*T
        u_ref(j,1)=mean_left;
    end

    if x(j)>= x0+s*T
        u_ref(j,1)= mean_right;
    end
end
```

C.2.4.4 exact_solution_p_inf.m

```
function [u_ref] = exact_solution_p_inf(mean_left,mean_right,sig_h,T,m,left,right,x0,x,p)

% Compute the analytical solution of the stochastic Burgers' equation with
% Hermite polynomials


% Indata:
% mean_left,mean_right - Mean (u_0) of the left and right states
% sig_h - Standard deviation (u_1), assumed uniform over the spatial domain
% T - Time
% m - Number of spatial grid points
% left - Lower limit of spatial interval
% right - Upper limit of spatial interval
% x0 - Initial shock location
% x - Vector of spatial grid points
% p - Number of gPC coefficients to be computed

% Outdata:
% u_ref - Analytical solution
```

```
y = zeros(m,1);
y(:,1)=(x-x0)./(sig_h*T)-(mean_left+mean_right)/(2*sig_h);

Psi_s = zeros(m,p);
Psi_s(:,1) = 1;
Psi_s(:,2) = y;


u_ref = zeros(m,p);
u_ref(:,1) = mean_left - (mean_left-mean_right)*normcdf(y,0,1);

if p>2
    u_ref(:,2) = sig_h + (mean_left-mean_right)*exp(-y.^2/2)/sqrt(2*pi);
    for k=3:p
        Psi_s(:,k) = y.*sqrt(factorial(k-2)/factorial(k-1)).*Psi_s(:,k-1) - (k-2)*sqrt(factorial(k-3)/
          factorial(k-1)).*Psi_s(:,k-2);
        u_ref(:,k) = (mean_left-mean_right)/sqrt(k-1)*exp(-y.^2/2)/sqrt(2*pi).*Psi_s(:,k-1);
    end
end
```

Index

M.P. Pettersson et al., *Polynomial Chaos Methods for Hyperbolic Partial Differential Equations*, Mathematical Engineering,
DOI 10.1007/978-3-319-10714-1

MIX
Papier aus verantwortungsvollen Quellen
Paper from responsible sources
FSC® C105338

If you have any concerns about our products,
you can contact us on
ProductSafety@springernature.com

In case Publisher is established outside the EU,
the EU authorized representative is:
Springer Nature Customer Service Center GmbH
Europaplatz 3, 69115 Heidelberg, Germany

Printed by Libri Plureos GmbH
in Hamburg, Germany